W9-ASH-593
4604 9100 023 007 8

The Scientific Revolution

Titles in the World History Series

The Age of Feudalism
The Age of Pericles
The American Frontier
The American Revolution
Ancient Greece
The Ancient Near East
Architecture
Aztec Civilization
Caesar's Conquest of Gaul
The Crusades
The Cuban Revolution
The Early Middle Ages
Egypt of the Pharaohs
Elizabethan England
The End of the Cold War
The French and Indian War
The French Revolution
The Glorious Revolution
The Great Depression
Greek and Roman Theater
Hitler's Reich
The Hundred Years' War
The Inquisition
The Italian Renaissance
The Late Middle Ages
The Lewis and Clark Expedition
Modern Japan
The Punic Wars
The Reformation
The Relocation of the North American Indian
The Roman Empire
The Roman Republic
The Russian Revolution
The Scientific Revolution
The Spread of Islam
Traditional Africa
Traditional Japan
The Travels of Marco Polo
The Wars of the Roses
Women's Suffrage

The Scientific Revolution

by
Harry Henderson
and Lisa Yount

Lucent Books, P.O. Box 289011, San Diego, CA 92198-9011

To all the young scientists
who will create the next revolution

Library of Congress Cataloging-in-Publication Data

Henderson, Harry, 1951-
The scientific revolution / by Harry Henderson and Lisa Yount
p. cm.—(World history series)
Includes bibliographical references and index.
Summary: Describes scientific discoveries that took place between 1550 and 1900, examines the development of the scientific method, and discusses the impact on people's views of their world.
ISBN 1-56006-283-5
1. Science—History—Juvenile literature. [1. Science—History.]
I. Yount, Lisa. II. Title. III. Series.
Q125.H5584 1996
509—dc20 95-51342
CIP
AC

Printed in the U.S.A.

Contents

Foreword

Each year on the first day of school, nearly every history teacher faces the task of explaining why his or her students should study history. One logical answer to this question is that exploring what happened in our past explains how the things we often take for granted—our customs, ideas, and institutions—came to be. As statesman and historian Winston Churchill put it, "Every nation or group of nations has its own tale to tell. Knowledge of the trials and struggles is necessary to all who would comprehend the problems, perils, challenges, and opportunities which confront us today." Thus, a study of history puts modern ideas and institutions in perspective. For example, though the founders of the United States were talented and creative thinkers, they clearly did not invent the concept of democracy. Instead, they adapted some democratic ideas that had originated in ancient Greece and with which the Romans, the British, and others had experimented. An exploration of these cultures, then, reveals their very real connection to us through institutions that continue to shape our daily lives.

Another reason often given for studying history is the idea that lessons exist in the past from which contemporary societies can benefit and learn. This idea, although controversial, has always been an intriguing one for historians. Those that agree that society can benefit from the past often quote philosopher George Santayana's famous statement, "Those who cannot remember the past are condemned to repeat it." Historians who ascribe to Santayana's philosophy believe that, for example, studying the events that led up to the major world wars or other significant historical events would allow society to chart a different and more favorable course in the future.

Just as difficult as convincing students to realize the importance of studying history is the search for useful and interesting supplementary materials that present historical events in a context that can be easily understood. The volumes in Lucent Books' World History Series attempt to present a broad, balanced, and penetrating view of the march of history. Ancient Egypt's important wars and rulers, for example, are presented against the rich and colorful backdrop of Egyptian religious, social, and cultural developments. The series engages the reader by enhancing historical events with these cultural contexts. For example, in *Ancient Greece*, the text covers the role of women in that society. Slavery is discussed in *The Roman Empire*, as well as how slaves earned their freedom. The numerous and varied aspects of everyday life in these and other societies are explored in each volume of the series. Additionally, the series covers the major political, cultural, and philosophical ideas as the torch of civilization is passed from ancient Mesopotamia and Egypt, through Greece, Rome, Medieval Europe, and other world cultures, to the modern day.

The material in the series is formatted in a thorough, precise, and organized manner. Each volume offers the reader a comprehensive and clearly written overview of an important historical event or period. The topic under discussion is placed in a broad historical context. For example, *The Italian Renaissance* begins with a discussion of the High Middle Ages and the loss of central control that allowed certain Italian cities to develop artistically. The book ends by looking forward to the Reformation and interpreting the societal changes that grew out of the Renaissance. Thus, students are not only involved in an historical era, but also enveloped by the events leading up to that era and the events following it.

One important and unique feature in the World History Series is the primary and secondary source quotations that richly supplement each volume. These quotes are useful in a number of ways. First, they allow students access to sources they would not normally be exposed to because of the difficulty and obscurity of the original source. The quotations range from interesting anecdotes to far-sighted cultural perspectives and are drawn from historical witnesses both past and present. Second, the quotes demonstrate how and where historians themselves derive their information on the past as they strive to reach a consensus on historical events. Lastly, all of the quotes are footnoted, familiarizing students with the citation process and allowing them to verify quotes and/or look up the original source if the quote piques their interest.

Finally, the books in the World History Series provide a detailed launching point for further research. Each book contains a bibliography specifically geared toward student research. A second, annotated bibliography introduces students to all the sources the author consulted when compiling the book. A chronology of important dates gives students an overview, at a glance, of the topic covered. Where applicable, a glossary of terms is included.

In short, the series is designed not only to acquaint readers with the basics of history, but also to make them aware that their lives are a part of an ongoing human saga. Perhaps they will then come to the same realization as famed historian Arnold Toynbee. In his monumental work, *A Study of History,* he wrote about becoming aware of history flowing through him in a mighty current and of his own life "welling like a wave in the flow of this vast tide."

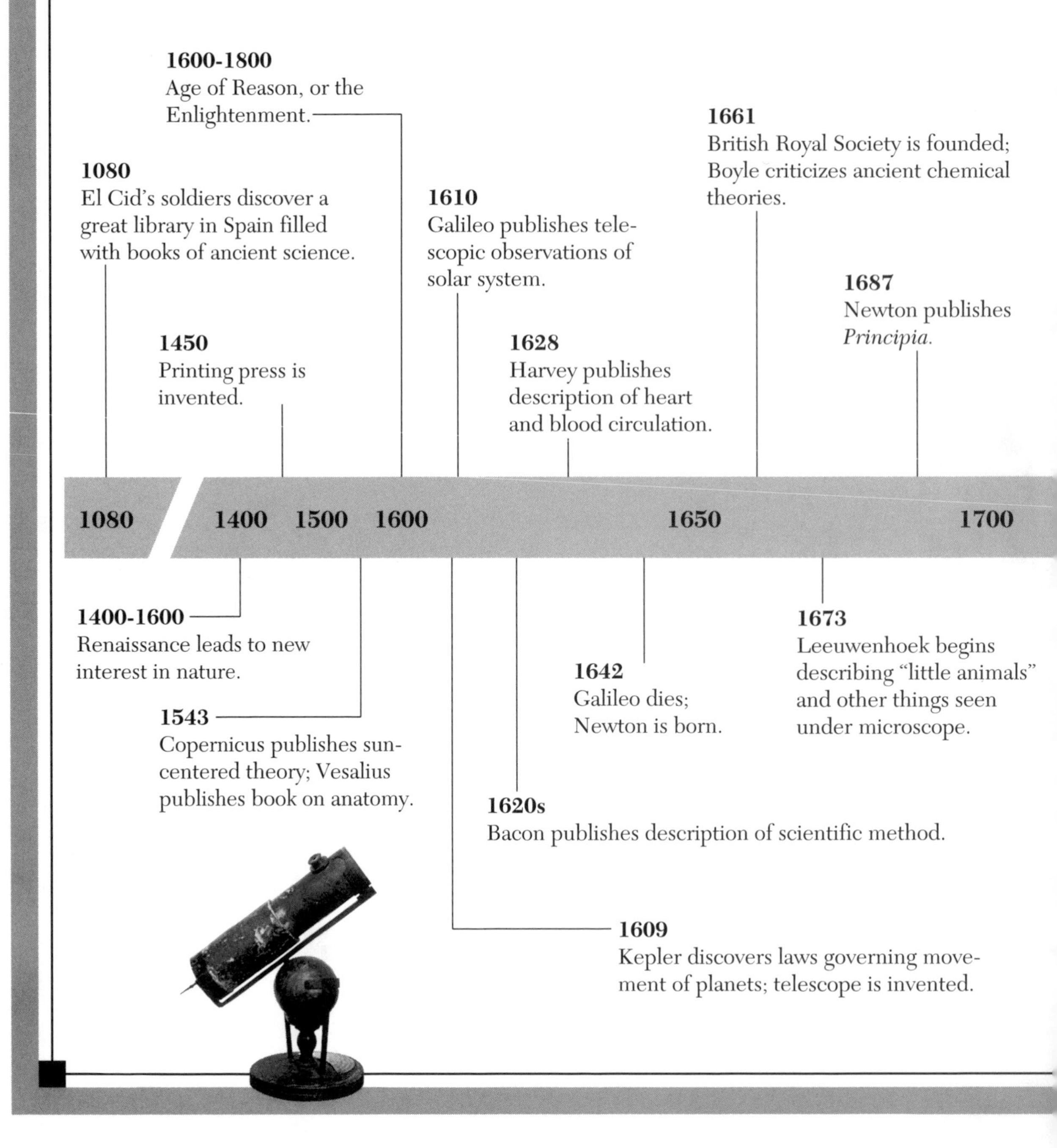
Important Dates in the History of the Scientific Revolution
1080
El Cid's soldiers discover a great library in Spain filled with books of ancient science.
1450
Printing press is invented.
1600-1800
Age of Reason, or the Enlightenment.
1610
Galileo publishes telescopic observations of solar system.
1628
Harvey publishes description of heart and blood circulation.
1661
British Royal Society is founded; Boyle criticizes ancient chemical theories.
1687
Newton publishes *Principia.*
1080
1400
1500
1600
1650
1700
1400-1600
Renaissance leads to new interest in nature.
1543
Copernicus publishes sun-centered theory; Vesalius publishes book on anatomy.
1620s
Bacon publishes description of scientific method.
1609
Kepler discovers laws governing movement of planets; telescope is invented.
1642
Galileo dies; Newton is born.
1673
Leeuwenhoek begins describing "little animals" and other things seen under microscope.

1800
Cuvier digs up bones of dinosaurs near Paris.

1869
Mendeleyev publishes periodic table of the elements; first dye is synthesized using principles of organic chemistry.

1859
Darwin publishes *Origin of Species.*

1831
Faraday discovers how to change magnetism into electricity.

1895
Roentgen discovers X rays; Becquerel discovers radioactivity.

1750 **1800** **1850** **1900**

1760
Franklin demonstrates lightning rod.

1808
Dalton publishes atomic theory.

1870s
Pasteur and Koch prove that microorganisms can cause human and animal diseases.

1789
Lavoisier publishes *Elements of Chemistry;* French Revolution begins.

1830
Lyell publishes *Principles of Geology.*

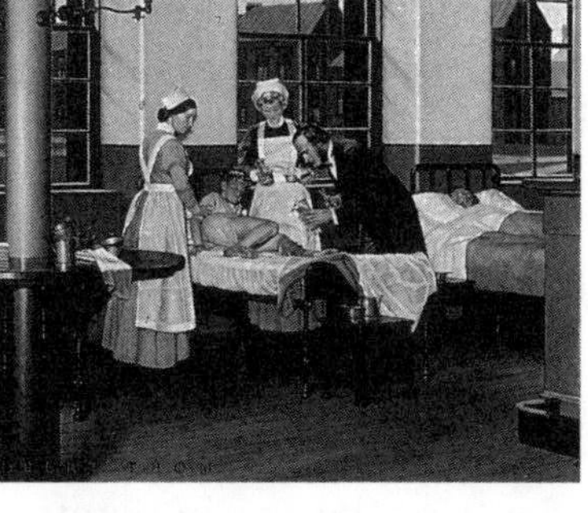

Three Stories of Science

The story of the scientific revolution is three exciting tales in one.

The first story is about the fascinating things that people, mostly in Europe, discovered about the world between about 1550 and 1900. Galileo Galilei looked at the sky through a telescope for the first time and saw mountains and valleys on the moon, a swarm of tiny moons circling the planet Jupiter, and hundreds of stars inside the Milky Way. Georges Cuvier dug into the hills around Paris and found the bones of woolly mammoths and great dinosaurs. Antonie van Leeuwenhoek peered into a microscope and saw a zoo of tiny creatures swimming in a drop of water. Within a little more than three centuries, world after new world of knowledge was revealed.

The second story is about how people learned how to become scientists. They began by making careful observations and measurements of the physical world. They also tried to explain what they saw and why it happened. Benjamin Franklin did this when he compared lightning to the force generated when a piece of amber, or fossilized tree sap, was rubbed with fur. Charles Darwin did it when he claimed that accidental variations in plants and animals, coupled with a scarcity of food and other resources in nature, brought about long-term changes that made living things better suited to their environment.

Scientific inquiry is different from other ways of explaining the world, such as religion or art, because science demands that each explanation be tested and proven true. One way to do this is by performing an experiment. An experiment is a test in which natural forces are allowed to operate under controlled conditions and the results are carefully observed and measured. For example, Benjamin Franklin was carrying out an experiment when he set up an electrical connection to the ground and then watched what happened as he drew lightning down from the sky. Another way to test an explanation is by collecting many observations over a long period of time. Darwin spent twenty years gathering examples, observed by himself and others, that supported his explanation of how plants and animals changed.

The third story is about how science changed people's way of seeing their world. Because of what Copernicus and Galileo started, people learned that the earth circles an ordinary star in a galaxy of millions. Because of what Isaac Newton discovered, people saw that the behavior of matter and energy could be explained by a few simple but powerful laws. Because of what Darwin

Galileo Galilei demonstrates his telescope to the Venetian Senate. Through this revolutionary invention, Galileo was able to gaze upon the valleys of the moon and the stars in the Milky Way.

began, they became aware that life on earth is a complex, ever-changing web of beings that are both fragile and amazingly resourceful. Science eliminated the ancient belief that human beings were the center of the universe, but it gave people the satisfaction of knowing that the human mind could comprehend many of nature's deepest mysteries.

Science has also changed the way people live. When doctors identified and found ways to kill disease-causing microorganisms, they banished diseases like smallpox and bubonic plague that killed or crippled millions. When scientists discovered that magnetism could be changed into electricity, they paved the way for inventors such as Morse, Bell, and Edison to create devices that helped people communicate better, entertained them, and lit their homes at the flick of a switch.

When you put all three stories together, you have the story of a new way of thinking that completely transformed the world in the space of less than four hundred years. It paved the way for the even more startling discoveries and sweeping changes that have occurred in the twentieth century and continue today.

The story of science has no end, only new beginnings. Perhaps someday you will add a new chapter to the story.

Chapter

1 The Birth of Modern Science

After the fall of the Roman Empire around A.D. 450, Europe gradually entered the time called the Middle Ages. The city life of the ancient world withered away. The new centers of power were the castles built by nobles and the churches and monasteries where bishops and abbots were in charge of spiritual life. The farming techniques of the time could barely feed the people, and little money was left

Peasants work the land surrounding their lord's castle. As cities and schools withered away during the early Middle Ages, castles became the new centers of power.

over for trade or expansion. Most people had to spend all their time and resources simply trying to survive. With few cities or schools, there was little knowledge of, or interest in, science. Most of the scientific discoveries of the ancient world had been forgotten.

Science as a Sin

Christianity dominated Europe in the Middle Ages, and it actively discouraged the pursuit of science. It taught that the secrets of nature were God's domain and that probing them was an intrusion into areas where humans were not meant to go.

Some quoted the Book of Job in the Bible, in which God demanded that the ancient patriarch Job answer these questions: "Where were you when I laid the foundations of the earth? If you have understanding, say so. Do you know who laid out its measures? Who has stretched a line upon it?"

God seemed to be telling Job that a mere human being was in no position to question him. The secrets of creation were forever beyond human understanding. The only thing that mattered was trying to live in a way that pleased God and would result in a place in heaven after death.

Many religious people felt that science might lead human beings away from God. The great church leader Augustine wrote:

> The good Christian should beware of mathematicians, and all those who make empty prophecies. The danger already exists that the mathematicians have a covenant [agreement] with the devil to darken the spirit and to confine man in the bonds of Hell.[1]

By mathematicians Augustine meant astrologers, who claimed to use their knowledge of the positions of the planets to predict the future. Augustine believed that people should not try to learn their future. Instead, they should simply have faith in God. He thought that trying to predict what will happen, which is a vital part of science, was not only useless but dangerous to the soul.

Awakening from the East

Some education still took place during the Middle Ages, of course. The church preserved some of the works of Aristotle and a few other Greek and Roman thinkers who had begun the systematic study of nature more than a thousand years earlier. Theologians, people who study God and religion, were taught skills of logical reasoning. But what led to the reawakening of science in Europe were new contacts between Europeans and a culture that was more advanced in science and technology.

Around the middle of the eleventh century, a king named Alfonso VI united three small Christian kingdoms in Spain. His armies, led by the famous warlord called El Cid, began a crusade to drive out the Arabs who had built an advanced civilization in Spain over several centuries.

El Cid's soldiers entered the Spanish city of Toledo in 1080. They could not believe what they saw: broad streets, fountains, and buildings more magnificent than anything that had existed in the rest of Europe since the time of the Romans. And soon news about a special treasure began to spread—great libraries filled with thousands of books. These books included translations of the works of Greek and Roman scholars as well as later works by Arab thinkers such as Avicenna, a physician and philosopher. The books covered subjects such as medicine, astronomy, chemistry, then called alchemy, geometry, and *al-jabr,* or algebra.

Miraculously, the invaders did not burn the books, as earlier Christians had done with the great ancient library in the Egyptian city of Alexandria. Instead, scholars began coming from all over Europe to study them. A community of Jewish scholars who had lived in Toledo for centuries aided the visiting scholars by translating the Arabic works into Latin so the newcomers could read them.

Further contact with Arabic culture came when the church began to sponsor crusades to seize control of parts of the Middle East, especially the city of Jerusalem, from their rulers—Arabs who followed the Muslim, or Islamic, religion. In the course of these crusades, Europeans gained access to trade routes that the Muslims had kept for themselves. Spices and other exotic goods from lands as far away as China began to flow into Europe along these routes. This flow of trade also exposed Europeans to a wealth of new science and technology from the Middle East and Far East.

New Centers of Learning

The growth of trade and a rising population in European towns and cities meant that more educated people were needed to manage Europe's thriving economy. It also meant that more resources were available for education. These changes led to the founding of universities in major cities, such as Paris in France and Bologna in Italy, in the eleventh and twelfth centuries. These great centers of higher education became fertile ground for the growth of science.

In 1122 one of the new university scholars, Peter Abelard, wrote a book called *Yes and No,* in which he described a system for determining whether a statement was true. Until that time people having a debate would support their cases by naming famous and respected persons who agreed with them. Abelard, however, said that scholars should allow for doubt and question everything. They should insist on the use of logical argument. Even the words of the Bible, Abelard said, were open to logical examination.

The idea that sound reasoning is more important than support by authorities underlies what has come to be called the scientific method. But when Abelard used his method to analyze statements from the Bible and said that they contradicted one another, he was bitterly attacked by the powerful church leader Bernard of Clairvaux. Bernard accused Abelard of heresy, or challenging accepted religious teachings. He ordered some of Abelard's books burned. Nevertheless, Abelard's methods were later adopted by scientific pioneers.

The Renaissance and Humanism

By the fifteenth century many changes were taking place in scholarship and in the arts. The rediscovery of Aristotle and other ancient thinkers led to a new interest in science. Architects, artists, and writers also

How to Solve a Problem

French philosopher René Descartes, in his Discourse on Method, *describes a procedure for solving scientific problems that is also good advice for test takers today. Descartes's procedure is summarized by Carole Collier Frick in* The Scientific Revolution.

"First of all, I have decided never to believe something is true unless I know it for certain.

Secondly, I will divide each thing which I do not understand about the problem into as many parts as possible so that I can begin to solve them one by one.

Third, I will start with the easiest and most simple part of the problems, and work through to the most difficult and complex parts, until I have solved the whole problem.

Lastly, I will keep complete notes of the method of investigation I have used and of the results I have reached, so that I am sure I have not forgotten anything."

turned to the ancient world for inspiration.

Because of the way ancient culture was revived during the fifteenth and sixteenth centuries, this period is often called the Renaissance, a French word meaning "rebirth." Greek and Roman culture had focused on the earthly world of nature and human society. In turning to and building on the works of Greece and Rome, Renaissance artists and thinkers went back to these worldly concerns. Their focus on human life, as opposed to God and religion, came to be called humanism.

Renaissance art was naturalistic, or true to nature. It showed the detailed structure of real things. Unlike the flat-looking paintings of the Middle Ages, Renaissance paintings appear solid and three-dimensional.

Artists in this period began to study anatomy, or the structure of the body. They learned how muscles give shape to the human body. Some artists also made accurate drawings of plants and animals as they appeared in nature. This type of drawing later proved useful to scientists who wished to draw or study pictures that showed the exact structure of living things.

Perhaps the most talented and remarkable person of the Renaissance was Leonardo da Vinci. Leonardo was an artist, scientist, and inventor. He did most of his work in the second half of the sixteenth century. He filled notebooks with drawings and writing in a peculiar script that ran across the page from right to left and had to be read using a mirror.

Leonardo was fascinated with the way things worked—or might be made to work. He studied birds' wings and used them as the basis for sketches of possible flying machines. He designed military forts and weapons. He made detailed

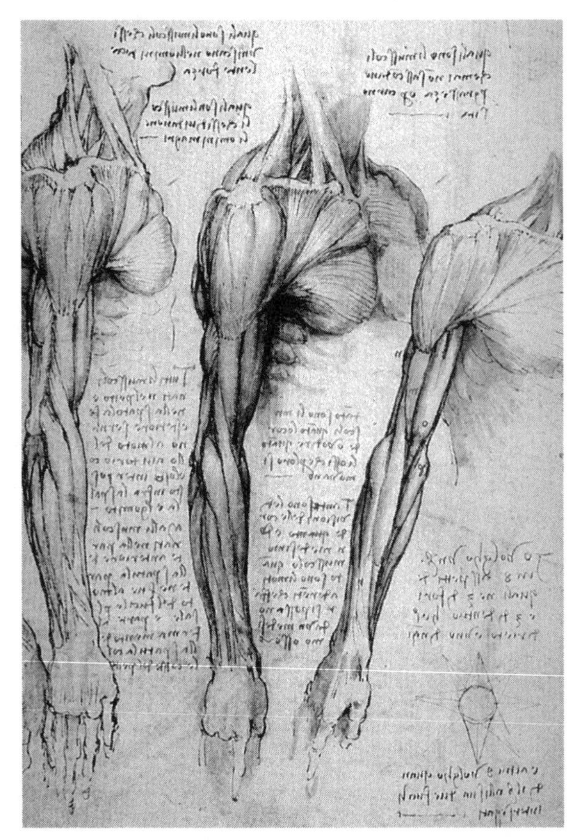

Renaissance artists were able to represent the body in a naturalistic style by studying anatomy. This drawing by Leonardo da Vinci illustrates the muscles in the arms, chest, and neck.

drawings of the structure and growth of plants, the flow of water, and the different layers of rocks in hillsides. In all his work Leonardo combined the eye and training of an artist with the careful observation and analysis of a scientist.

Leonardo worked in secret, fearing misunderstanding, and some of his notebooks were not discovered until our own century. He therefore had little direct impact on the development of science during the Renaissance, although he was greatly respected as an artist. Nonetheless, his work is a fine example of the growing curiosity about nature among Europeans. Some of it also foreshadowed the science and technology of later centuries.

Observation and New Instruments

The rediscovery of ancient science proved to be a mixed blessing to the new scientists of Europe. Writers such as Aristotle, in physics, and Ptolemy, in astronomy, opened up a vast new world of possibilities for studying nature. But the work of the ancients was very advanced compared to

Leonardo da Vinci was an accomplished artist, scientist, and inventor. He filled numerous notebooks with detailed sketches and insightful observations about the laws of nature.

Galileo lets curious bystanders observe the satellites of Jupiter through his telescope, an invention that revolutionized the seventeenth-century world.

what Europeans of the Middle Ages had known. The scientists of the Renaissance, therefore, tended to accept it without question.

Ancient writers quickly became the authorities in every scientific field. Doubting their ideas became almost as frowned upon as doubting the Bible. Abelard's advice to "question everything" was all but forgotten. As a result, Renaissance scientists at first seldom went beyond what the ancients had written.

Ironically, some of the ancient writers themselves would have hated this attitude. They had stressed the importance of independent thought and direct observation. For example, Galen, a second-century Roman doctor whose works became the most respected source of information on medicine during the Renaissance, had written, "The surest judge of all will be experience alone."[2] But when Renaissance doctors found that the structure of real human bodies sometimes did not match the descriptions given by Galen, they tended to ignore the differences.

In order to make discoveries through observation that went substantially beyond ancient science, Renaissance scientists needed to be able to observe with more than their five senses. The ancients had had few scientific instruments—mainly measuring rods and simple devices for finding the positions of stars or planets in the sky. Starting around the seventeenth century, however, new tools revolutionized the way scientists worked.

The invention of the telescope made it possible for an astronomer named Galileo to see things about the solar system that no ancient astronomer could have observed. Galileo also invented a thermometer to measure temperature. A later instrument called a barometer made it possible to study weather and the behavior of gases. The microscope revealed a hidden world of tiny living things.

These new instruments allowed scientists to observe nature in much greater detail than before. By revealing things that were unknown to the ancients, the instruments encouraged scientists to question authority. Gradually scientists realized that the best answers to their questions would come, not from what someone once said, but from what they saw with their own eyes.

Explanations and Experiments

Scientists also came to realize that they would need to go beyond simple observation, even with improved instruments, if they wanted to learn nature's deepest secrets. They would need to experiment—to make small, planned changes in some part of nature and observe the results. Experiments are vital to science because they help scientists explain *why* nature behaves the way it does.

Sometimes observation is enough to show whether a proposed explanation, or hypothesis, for something in nature is correct. For example, a scientist might work out a mathematical formula that predicts how a planet will move. The scientist can check the predictions by measuring the planet's actual movements and comparing them with those predicted by the formula.

In many cases, however, experiments are the best way to test a hypothesis. For example, Aristotle had written that heavy objects fall faster than light objects. According to legend, Galileo decided to perform an experiment to find out whether this was true. He climbed to the top of a tall tower and dropped cannonballs of different weights from it. He found that Aristotle was wrong: the balls hit the ground at the same time, regardless of their weight. This experiment let Galileo answer his question much more quickly and effectively than he could have done by simply observing objects that happened to fall when he was nearby.

Galileo proves that all objects, regardless of weight, fall at the same rate. This legendary experiment refuted the current Aristotelian theories about gravity.

Some scientists had recognized the importance of experiments even before the Renaissance. A thirteenth-century English thinker named Roger Bacon insisted that "the strongest argument proves nothing so long as the conclusions are unverified [untested] by experience. Experimental science is the queen of science."[3] Bacon himself carried out many scientific experiments. His ideas were not widely accepted at the time, however. He was accused of sorcery, or black magic, and put in prison for years.

The Scientific Method

Early in the seventeenth century Francis Bacon, unrelated to Roger, complained:

> Our knowledge of science has not progressed at all, and the same old questions remain unanswered. Instead of discussing ideas, and coming up with new answers to problems, our schools are just teaching the same old things.
>
> There are a few people who do conduct experiments and invent things, but when they do, they don't follow a standard step-by-step procedure in their work or a well-planned overall strategy. . . .
>
> What I would like to do is to suggest a new method people can follow when they want to work on an important problem. This way, our knowledge of science will progress.
>
> I believe people should restrict themselves to recording the results of carefully designed experiments. The experiments themselves will show whether a hypothesis is true or false. Those scientists who don't want to be just guessing, but really to be able to discover and know things, must rely not on someone else's previous conclusions, but on their own factual evidence for everything.[4]

The approach to studying nature that Bacon outlined in his book *Instauratio Magna* (*The Great Restoration*) is now called the scientific method. It is at the heart of the revolution in science that began in the Renaissance and still goes on today. In the scientific method a scientist uses imagination to picture what *might be* and then tests those ideas by comparing them to what *is*. Basically, the method consists of these four steps: 1) making a hypothesis; 2) designing and carrying out experiments to test the hypothesis; 3) using instruments to make precise measurements of what occurs during the experiments; 4) revising the hypothesis to reflect what the experiments reveal.

Checking one's own work and that of others by repeating experiments is also part of the scientific method, so sharing information became essential to the growth of science. The printing press, invented in the middle of the fifteenth century, helped scientists spread information in written form. Beginning in the seventeenth century, scientists also set up groups, or societies, that met to hear and discuss reports of their members' work. The best-known scientific societies included the Royal Society in Great Britain and the Academy of Sciences in France.

Science and Progress

The growth of science and the new focus on human life and nature during the Renaissance gave rise to the idea of progress. In the Middle Ages most people did not try to invent or discover new things. By 1600, however, people could already see that new technology growing out of scientific discoveries was changing and often improving their world. They began to hope that science would provide more improvements as time went on. They came to see history as a process of moving forward and upward from simple beginnings.

In 1620 Francis Bacon wrote a book called "*Novum Organum,* or 'The New Instrument.'" He intended this book to be a modern replacement for a similarly titled book by Aristotle. In it he noted:

Science by Teamwork

Francis Bacon, in The New Atlantis, *described the workings of an ideal scientific institute. His ideas helped inspire the foundation of such organizations as England's Royal Society. This quote is taken from Carole Collier Frick's book* The Scientific Revolution.

"1. We have twelve people who sail into foreign countries, who bring us the books, outlines, and patterns of experiments from all other parts [of the world].
2. We have three people who collect all the experiments which are in all books.
3. We have three people who collect all the experiments of all mechanical arts, and also of liberal [theoretical] sciences, and also of other practices.
4. We have three people who try new experiments, such as they think good.
5. We have three people who organize the experiments of the former four [groups] into titles and tables, to help us understand them.
6. We have three people who analyze the experiments of their fellows, and try to figure out useful applications for our lives.
7. After different meetings and consultations of our whole number to consider these above-mentioned activities, we have three people who direct new, more advanced experiments, which are more penetrating into nature than the former.
8. We have three others that carry out the experiments so directed, and then report to us on them.
9. Lastly, we have three people who make larger generalizations from the above experiments."

Francis Bacon called on scientists to use a standard method to test their theories. His scientific method emphasized the need for scientists to conduct experiments, precisely record data, and then modify their hypotheses to include the experimental findings.

> It is well to observe the force and virtue [power] and consequence of discoveries; and these are to be seen nowhere more conspicuously than in those three which were unknown to the ancients . . . ; namely, printing, gunpowder, and the magnet. For these three have changed the whole face and state of things throughout the world; the first in literature, the second in warfare, and the third in navigation; whence [from which] have followed innumerable changes; insomuch that no empire, no sect [religion], no star seems to have exerted more power and influence in human affairs than these mechanical discoveries.[5]

Actually, all three of the inventions Bacon mentioned had existed for centuries in China—something Europeans did not like to admit. But even if he was wrong about the age of the inventions, Bacon correctly reflected Europeans' feelings about them.

Bacon's words showed that by his time people were beginning to see science as more than just a way to learn about nature. They were realizing that it was also a way to put knowledge to work in inventions that *changed* nature. Through technology, science promised to bring a better life to everyone.

Chapter

2 The Workings of the Universe

In 1633, an elderly scientist sat before a row of red-robed church judges. Finally he said the words that would keep him from prison and perhaps torture:

> I, Galileo Galilei, son of the late Vincenzio Galilei, of Florence, aged 70 years, being brought to judgment . . . abandon the false opinion which maintains that the Sun is the center [of the universe] and immovable, and I will not hold, defend, or teach the said false doctrine in any manner.[6]

Although it seemed the Church had silenced him, Galileo is said to have muttered "and yet it [the earth] moves!" as he was led away from the court. No matter

Galileo stands before the Inquisition. In 1633, he renounced his controversial beliefs after increased pressure from church officials.

what he had been forced to say, he still believed that the earth traveled around an immovable sun, rather than the other way around.

No court could stop people from looking through a new invention, the telescope, and seeing the true nature of the solar system with their own eyes. In little more than a century, astronomers would present an astonishing new picture of the universe, and scientists in the new field of physics would explain why the universe worked the way it did.

New Views of the Universe

One of the books the crusader knights of the eleventh and twelfth centuries had found in the Arab libraries of Spain was called the *Almagest,* which means "The Greatest." It had been written by the ancient Egyptian astronomer Ptolemy in the second century. Ptolemy had brought together the best findings of the astronomers of his time.

The ancient astronomers had come to three fundamental conclusions about the universe: the sun and planets moved around the earth, the stars were fixed and unchanging, and the heavenly bodies—planets and stars—were perfect spheres that had no blemish or irregularity. Between the middle of the sixteenth century and the beginning of the seventeenth all of these conclusions were shown to be wrong.

The first challenge was aimed at Ptolemy's ideas about the motion of the planets. The planetary observations in the *Almagest* were reasonably accurate, but they could not quite account for some of the ways planets moved.

The ancient astronomer Ptolemy believed that the earth was the center of the universe, as illustrated in this elaborate woodcut.

The word *planet* means "wanderer" in Greek, and it was a good name for these mysterious heavenly bodies. While stars wheeled across the sky in regular paths like clockwork, the planets seemed to speed up and slow down in their motion. What was even more confusing, some planets, such as Mars, actually appeared to stop and then move backwards for a while.

Ptolemy had assumed that, wander though they might, the planets basically circled around the earth, as did all the other heavenly bodies. Such a view seemed to be only common sense. After

Nicolaus Copernicus believed that the sun was the center of the universe; however, he feared persecution and did not publish his theory until just before his death.

all, the earth *feels* motionless. The sun, moon, and stars can be seen to move across the sky. The Bible and the ancients both had agreed that the earth was the unmoving center of the universe.

Almost no one questioned this view until around the middle of the sixteenth century, when a Polish doctor named Nicolaus Copernicus happened to work as a canon, one of the officials in charge of a cathedral, in the town of Frauenburg. Copernicus spent much of his spare time high in one of the towers surrounding the Frauenburg church, making astronomical observations. He became especially interested in the problem of the wandering planets.

Ptolemy and later astronomers had tried to solve the problem of the planets' wayward motion by saying that each planet, in addition to its path around the earth, moved in its own small circle, called an epicycle. When a planet appeared to change speeds or go backwards as seen from the earth, this meant that the planet was simply changing directions in its own circle as it continued to move around the earth. Gradually, astronomers had drawn more and more complicated systems of circles in an attempt to predict the motions of planets more accurately. Nevertheless, the planets still seemed to wander away from their calculated positions.

This confusing system bothered Copernicus. He wondered if there might be a simpler answer. Suddenly he realized that the understanding of the planetary system could be simplified greatly if he changed just one basic assumption. If he imagined that the sun rather than the earth was the center of the system, most of the motions of the planets could be explained much more easily than they were in Ptolemy's account. For example, the seeming retrograde, or backward, motion of some planets could be seen as the result of the earth's moving ahead of them in their respective paths, or orbits, around the sun. There would be no need to imagine epicycles.

Copernicus did not dare to write down his new heliocentric, or sun-centered, theory at first. He feared that it would be ridiculed or, worse, would arouse the anger of the church. In those days church officials could severely punish or even execute people who disagreed with the Bible or with religious teachings. But in 1543, just before his death, Copernicus finally gave in to the urgings of his students and allowed his book, *On the Revolutions of the Heavenly Spheres,* to be published. Perhaps to soften the opposition he expected from the church, he wrote an introduction dedicating his book to Pope Paul III. In it, he noted:

> I hesitated for a long time as to whether I should publish that which I have written to demonstrate the earth's motion, or whether it would not be better to follow the example of the Pythagoreans [an ancient sect], who used to hand down the secrets of their philosophy to their relatives and friends in oral [spoken] form. . . . I was almost impelled [forced] to put the finished work wholly aside, through the scorn I had reason to anticipate on account of the newness and apparent contrariness to reason of my theory.[7]

Copernicus's fears about how his theory would be received proved correct. Indeed, one of the few things Protestant leaders Martin Luther and John Calvin

A Perfect Sphere?

In On the Revolutions of the Heavenly Spheres, *Copernicus explains why he believes the universe must be built upon perfect spheres. This excerpt appears in* A Treasury of World Science.

"First of all, we assert that the universe is spherical; partly because this form, being a complete whole, needing no joints, is the most perfect of all; partly because it constitutes the most spacious form, which is thus best suited to contain and retain all things; or also because all discrete [separate] parts of the world, I mean the sun, the moon, and the planets, appear as spheres; or because all things tend to assume the spherical shape, a fact which appears in a drop of water and in all fluid [liquid] bodies when they seek of their own accord to limit themselves. Therefore no one will doubt that this form is natural for the heavenly bodies."

and the officials of the Catholic Church agreed on was that Copernicus's book was a dangerous attack on religious truth. In 1615 the church put Copernicus's book on a list of books that Catholics were forbidden to read. It stayed there for over two centuries.

Copernicus's new theory was not quite correct in its description of how the planets moved. Nevertheless, his book on planetary motion was revolutionary indeed. A great twentieth-century scientist, Harold C. Urey, considers Copernicus to be the founder of modern science:

> All superlatives fail when describing the work of Nicolaus Copernicus. He broke with a conception of the solar system that had stood for one thousand years, and introduced an entirely new concept of the relation of the planets to the sun. In so doing he initiated the whole modern method of scientific thought, and modified our thinking on all phases of human life.[8]

A New Star

Renaissance astronomers also looked beyond the planets to the more distant stars. Since the time of the ancients, astronomers had believed that the stars were fixed and unchanging points on a transparent sphere that covered the earth like an overturned bowl. But in 1572 a startling event called this idea into question.

One evening in November of that year, a Danish astronomer named Tycho Brahe was looking at the sky just after sunset. Brahe knew the night sky very well. He had built measuring instruments more accurate than any used before and over the years had painstakingly plotted the positions of the stars and the paths of the planets. Like Copernicus, he had found that Ptolemy had made important mistakes. Brahe did not entirely accept Copernicus's ideas, however. On this evening, Brahe noticed that

> a new and unusual star, surpassing the other stars in brilliancy, was shining almost directly over my head; and since I had, almost from my boyhood, known all the stars of the heavens perfectly . . . it was quite evident to me that there had never before been any star in that place in the sky, even the smallest, to say nothing of a star so conspicuously bright as this.[9]

In 1572, Tycho Brahe discovered a new star in the evening sky. The revelation that the heavens were not unchanging challenged ancient wisdom and religious faith.

Brahe was not the only one who noticed the new star. Since both ancient Greek philosophers and later Christians saw the stars as a symbol of eternity and divine perfection, this unexplained star was disturbing to everyone who recognized it. Stars were never supposed to change. Where, then, could a new star come from? If new stars could suddenly appear, perhaps the universe did change, after all. Indeed, astronomers would soon find that this was very much the case. The second conclusion of the ancient astronomers, like their conclusion about the motion of the sun and planets, had to be wrong.

Ancient astronomers' third belief, that heavenly bodies were perfect in form and flawless in appearance, would soon come into question as well. To disprove this idea, Renaissance astronomers would call on a new invention, the telescope.

The telescope constructed by Galileo Galilei. With this invention, Galileo was able to examine the moon as well as the planets Venus and Jupiter.

Galileo and the Telescope

In 1609, about forty years after Brahe had described his new star, an Italian mathematician and astronomer named Galileo Galilei noted a rumor that "an optical instrument had been [invented] by a Dutchman, Johannes Lippershey, by the aid of which visible objects, even though far distant from the eye of the observer, were distinctly seen as if near at hand."[10] The Dutch were excited about this invention, the telescope, because they thought it would help them spot enemy ships while the ships were still far away at sea.

The new invention interested Galileo for another reason. He wondered what would happen if he pointed a telescope at the sky. After some tinkering Galileo succeeded in building his own telescope. It was less powerful than many binoculars that can be bought today—but it was powerful enough to show startling things about the heavens.

In 1610 Galileo published a book called *The Starry Messenger*, in which he reported what he saw through his telescope. It was a kind of diary with sketches. One of the first things he looked at was the moon. To the unaided eye, the moon looks like a kind of shiny ball, with a few dark areas that look like a face to some people. Looking through his telescope, however, Galileo reported "that the moon is not robed in a smooth and polished surface, but is in fact rough and uneven, covered everywhere, just like the earth's surface, with huge prominences [mountains], deep valleys and chasms [gorges]."[11]

As with Brahe's new star, Galileo's observations of the moon challenged both ancient wisdom and religious faith. The

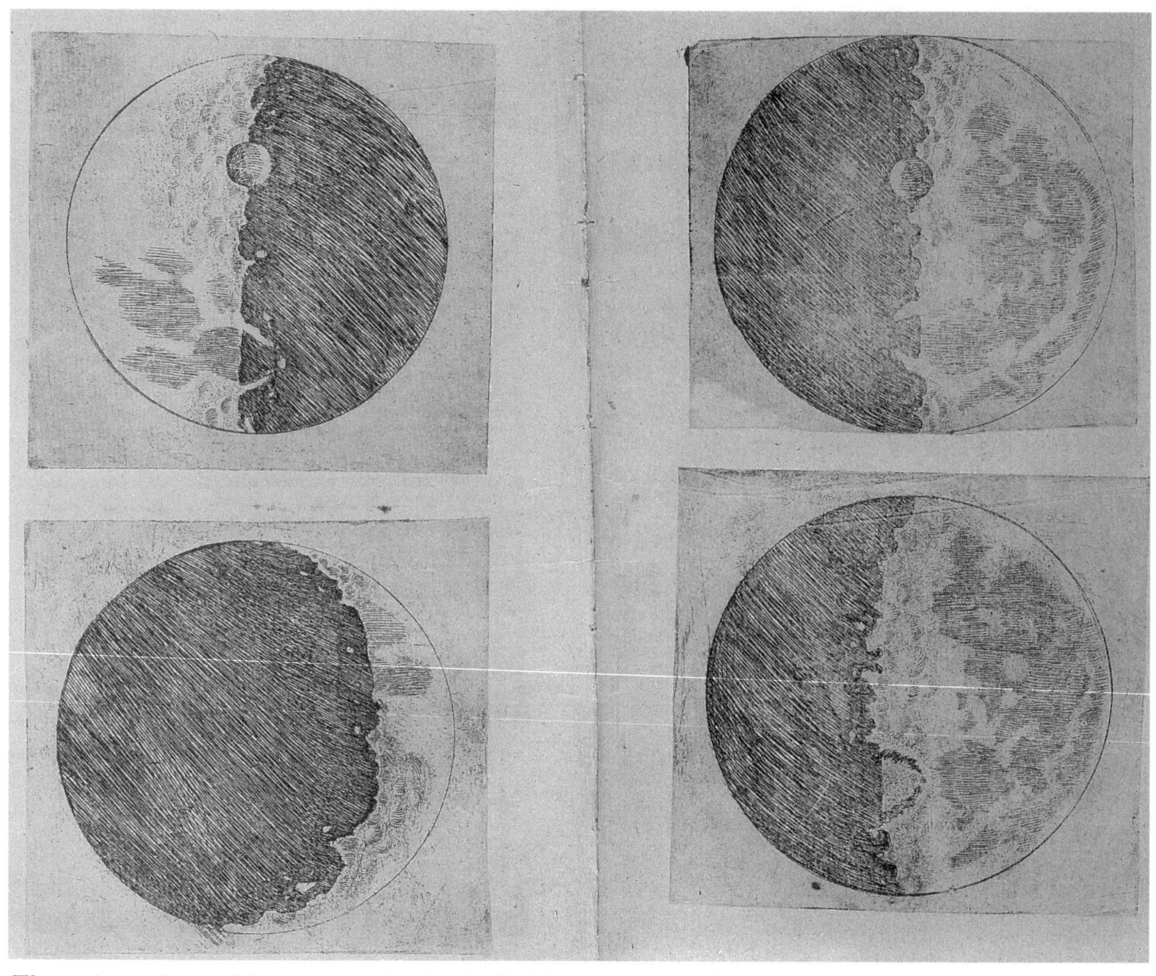

The various phases of the moon, as sketched by Galileo.

ancients had taught that the heavenly bodies were perfect spheres made of an eternally unchanging substance. Church scholars claimed that the heavens, being nearer to God than our world, shared in God's perfection. Galileo's telescope, however, revealed that the moon was a world with landscapes like those found on earth. It was anything but perfect.

When he turned his telescope to the planet Venus, Galileo saw that Venus, like the moon, showed phases. Over a period of time it appeared to wax and wane, that is, grow larger and smaller and change shape. And when he looked at Jupiter, he saw that the planet was accompanied by several tiny bodies that changed position over time. Clearly these bodies were moving around Jupiter, just as our own moon moves around the earth. This observation convinced Galileo that Copernicus's idea that the earth was not the center of the solar system was correct.

Galileo made his agreement with Copernicus obvious in the book *A Dialog Concerning Two World Systems*, which was published in 1630. This book consists of an imaginary conversation in which scholars debate whether Ptolemy or Copernicus is right about the motion of the planets. Although Galileo did not say which system he supported, his discussion shows the Copernican system in the better light.

Galileo knew that church officials would react violently to his writings, just as they had to the ideas of Copernicus. Galileo hoped he could convince church leaders that science, which discovered truth through observation as he had done with the planets, need not be a challenge to faith. In a letter to the Grand Duchess Christina of Tuscany, he suggested that science and religion were separate areas and that the Bible should not be used to judge scientific questions:

> I think that in discussions of physical problems we ought to begin not from the authority of scriptural passages, but from sense-experiences and necessary demonstrations; for the holy Bible and the phenomena of nature proceed alike from the divine Word. . . . For that reason it appears that nothing physical which sense-experience sets before our eyes . . . ought to be called in question (much less condemned) upon the testimony of biblical passages which may have some different meaning beneath their words.[12]

The Catholic Church did not accept Galileo's argument. In 1633 church leaders put him on trial for heresy, convicted him, and forced him to recant, or publicly deny, his beliefs. Even after he did so, he

Sunspots and Seeing Things Differently

In a letter about sunspots (which are now known to be magnetic disturbances on the sun), Galileo explains why ways of thinking have to change when new evidence is discovered. Galileo's letter, translated by Stillman Drake, is quoted in Frick's The Scientific Revolution.

"First of all, I have no doubt that they [the sunspots] are real objects and not mere appearances or illusions of the eye. . . . I have observed them for about 18 months, having shown them to various friends of mine. . . . It is also true that the spots do not remain stationary upon the body of the sun, but appear to move in relation to it with regular motions. . . . The spots seen at sunset seem to change place from one evening to the next. . . .

It proves nothing to say . . . that it is unbelievable for dark spots to exist in the sun simply because the sun is a most lucid [clear and bright] body. So long as men were obliged to call the sun 'most pure and most lucid,' no shadows or impurities whatever had been perceived in it; but now that it shows itself to us as partly impure and spotty, why should we not call it 'spotty and not pure'?"

had to live under a kind of house arrest. But the world of the seventeenth century was not that of the Middle Ages. Christians were now divided into Catholics and Protestants, and the Catholic Church was no longer the unquestioned ruler of Europe. Scientists could continue their work in countries where local rulers protected them. Even Galileo was able to do some scientific work after his trial.

The church stopped Galileo's work in astronomy, but it could not suppress what he had discovered—or the method he used. Galileo and his telescope had revealed things about nature that no one had seen before. The "evidence of the senses," aided by new technology, would be the foundation of science from then on.

Kepler and Planetary Motion

About fifteen years before Galileo turned his telescope to the sky, a German astronomer and mathematician named Johannes Kepler took up the question of how to predict the motions of the planets more accurately. Like Galileo, Kepler had become convinced that Copernicus's sun-centered planetary system was basically correct. Kepler found, however, that putting the sun rather than the earth in the center of the solar system was not enough to produce truly accurate predictions of the planets' movement. In 1596, at the age of twenty-four, Kepler published a book called *Cosmographic Mystery* in which he declared his support for the Copernican system. Kepler shared the belief of the ancient Greek school of philosophy called Pythagoreanism that "all nature is geometry," and his book attempted to use many-sided geometric figures called polygons to plot the orbits of the planets.

Tycho Brahe was so impressed with Kepler's book that he invited the young astronomer to be his assistant. When Brahe died a year later, Kepler inherited his notebooks, which were filled with the most accurate observations of planetary positions available at the time. Kepler turned his attention to the data about Mars because that planet had the most interesting motion, with its occasional backwards loops.

When Kepler tried to find a geometrical figure that matched the positions of Mars, he discovered that neither Copernicus's circles nor the polygons from his own book fitted the red planet's wayward

Johannes Kepler, an astronomer and mathematician, speculated that the planets move in elliptical orbits around the sun.

path. Instead, Kepler found that Mars moved in a path shaped like an ellipse. An ellipse is like an egg shape, or oval. While a circle has only a single center, an ellipse has two centers, or foci; the farther apart the foci are, the flatter and less like a circle an ellipse will be.

Later Kepler found that the other planets in the solar system also moved around the sun in orbits shaped like ellipses. In a series of new books, Kepler announced three laws, or mathematical rules, that together accounted for the motion of all the planets in the solar system. If scientists made accurate observations of a planet's present position, they could use Kepler's laws to predict its position for any future time.

The discoveries of Johannes Kepler and Galileo Galilei greatly challenged the ancient teachings of Aristotle (pictured).

Laws of Motion

Kepler showed *how* the planets moved around the sun. But *why* did objects in the universe move the way they do? This question intrigued Kepler's contemporary, Galileo. In the centuries following the Renaissance, researchers in a new branch of science would devote even more attention to it.

If the bodies in the heavens were part of nature and followed natural laws, these scientists wondered, might other laws describe how objects move on earth? Indeed, might objects in space and on earth follow the *same* natural laws?

The attempt to understand how objects move was part of the science now called physics. In the seventeenth century it was called natural philosophy. Physics later came to include the study of the nature of energy, such as light and heat.

When Renaissance scientists like Galileo and Kepler went to school, they had been taught the view of the physical world laid out by the ancient Greek thinker Aristotle. Aristotle had said that every kind of thing had its proper place in the universe. He explained that rocks fell to earth because they were trying to get back to where they belonged. Water flowed downhill because it was trying to go back to the ocean. Aristotle did not think that there was any one law that explained how everything—rocks, water, planets in the sky—moved. Rather, each kind of thing behaved according to its own tendency, or nature.

Kepler's discovery of laws of planetary motion, however, suggested that what objects were made of did not matter. The

Scientist Christian Huygens constructed a pendulum clock in 1656. Here, he presents the newly fashioned timepiece to King Louis XIV.

important things in determining motion were where objects were in relation to one another and how different forces acted on them.

Galileo, the Pendulum, and Acceleration

Galileo's interest in motion seems to have begun in his college days. One day during that time, so legend has it, he became bored during a long church service. Watching a chandelier, an ornate overhead light fixture, swinging back and forth, he used his heartbeat as a kind of clock to measure how much time it took for each complete swing. The swinging chandelier was acting as a pendulum, or object suspended so that it can move back and forth freely. Galileo discovered that a pendulum gradually slows down as it swings. The distance it covers in each swing grows shorter, but it still takes the same amount of time for each swing.

Since a pendulum thus kept steady time, Galileo suggested that medical students use it to measure how fast their patients' hearts beat. Knowing the speed of the heartbeat, or pulse, would help them identify some kinds of illness. He also suggested that a pendulum be used as part of the timekeeping mechanism of a clock.

Another scientist, Christian Huygens, built just such a clock in 1656. Tall grandfather clocks still use a pendulum.

In his later experiments Galileo used a water clock (which worked by dripping water into a bucket at a regular rate) to measure how falling objects sped up, or accelerated. He rolled balls down an inclined wooden beam, first letting a ball roll the full length, then half the length, a quarter, and so on. By weighing the water in the clock bucket, he could measure how much time the ball took to roll each distance. When he looked at hundreds of such measurements, he found that the distance increased with the square of the time. In other words, if a ball rolled five feet in one time period, it would roll (5 X 2 X 2), 20 feet in two periods, (5 X 3 X 3), 45 feet in three periods, and so on.

Galileo put his results to practical use by plotting the path of a cannonball. He broke the path into two parts: a horizontal line from the cannon to the target and a vertical line that showed the rate at which the ball fell toward the ground under the pull of gravity as it flew. Plotting the cannonball's position on these two lines at different times produced a curve that represented the real path of the ball. Using Galileo's formulas, armies made tables that told gunners the angle at which to aim a cannon so that the balls would hit a target at a given distance.

After church authorities ended Galileo's career in astronomy in 1633, he returned in his last years to the seemingly safer subject of motion on earth. In 1638 he published a book called *Discourses and Mathematical Demonstrations on Two New Sciences*, which summarized his lifelong research on falling bodies, motion, and the pendulum. This work was as important in establishing a modern science of physics as his observations of the heavens were in establishing modern astronomy. Galileo was one of the earliest scientists to show that experiment, measurement, and mathematics could be used to create precise descriptions of how objects behaved on earth.

Newton's New Mathematics

A few months after Galileo died in 1642, Isaac Newton was born in the English countryside at a village called Woolsthorpe. His father had died two months before his birth, and Newton was raised by a grandmother. He was a rather lonely, dreamy boy, and it soon became clear that he would not make a very good farmer.

Isaac Newton developed the branch of mathematics known as calculus to help scientists compute complex formulas such as the areas covered by planetary motion.

Instead, he was sent to Cambridge University in 1661.

Four years later, his less than outstanding college career was interrupted by an epidemic of bubonic plague, or Black Death, which caused him to be sent home. There he had nothing to do but concentrate on mathematics and science.

Earlier scientists such as Galileo and Kepler had used simple geometry and algebra to work out formulas for describing the motion of falling bodies or whirling planets. Newton, however, realized that advancing further in physics required a kind of mathematics that could express how one quantity, such as acceleration, varied in relation to another, such as time. He developed a new branch of mathematics, which today is called calculus. With calculus Newton could deal with rates of change and calculate the areas covered by curved paths such as the motion of planets or falling bodies.

Universal Gravitation

While developing calculus, Newton began to think about gravity, the force that makes objects fall to earth. He later recalled:

> In those days, I was in the prime of my age [best part of my life] for invention and minded [paid attention to] mathematics and philosophy [physics] more than at any time since. . . . I began to think of gravity extending to the [orbit] of the moon . . . , and from Kepler's rule of the periodical time of the planets being [in] proportion to the [square of] their distances [from the sun], I deduced that the forces that kept the planets in their [orbits] must be as [one divided by] the squares of their distances from their centers about which they revolve. . . . [I] compared the force [necessary] to keep the moon in her [orbit] with the force of gravity at the surface of the earth, and found them to answer [agree] pretty nearly.[13]

In other words, Newton realized that the same force Kepler had seen at work in the motions of the planets around the sun must also be responsible for the moon's rotating around the earth. As the distance of the orbiting object—moon or planet—from the object it went around doubled, the strength of the force holding it in orbit fell by one-fourth. If the distance were three times as great, the force would be one-ninth, and so on.

Newton had discovered the universal law of gravitation: that every body in the universe attracts every other body with a force proportional to the number one divided by the square of the distance between the objects. (Legend has it that Newton was inspired to discover the law of gravitation by watching an apple fall from a tree in his garden. It's a nice story, and it may even be true.)

Newtonian Mechanics

The next question was this: how does gravity make the planets move in the elliptical paths that Kepler had observed? In 1684 two English scientists, Edmund Halley and Robert Hooke, sat down to lunch in a London tavern with the famous architect Christopher Wren. They compared notes

An artist's rendition depicts Isaac Newton observing apples falling from a tree, the experience that is believed to have inspired his discovery of the universal law of gravitation.

about the law of gravitation, which the two scientists had discovered on their own, for Newton, who tended to be secretive, had not published his findings. Halley felt that there should be a mathematical way to show why a force of gravity that fell as the number one divided by the square of the distance would produce an elliptical orbit. Wren, who was not a scientist but who had served as president of the Royal Society, offered an expensive book as a prize to whichever of his two scientist friends could provide a mathematical proof.

After working on the problem for two months without success, Halley decided that Newton might be able to help him. When the great comet of 1680 had appeared, Newton had asked Halley for the data from his observations of that body. Now Halley hoped Newton would return the favor.

> The Doctor [Halley] asked [Newton] what he thought the curve would be that would be described [traced out] by the planets, supposing that the

force of attraction toward the Sun to be reciprocal to the square of their distance from it. Sir Isaac replied immediately that it would be an ellipse. The Doctor, struck with joy and amazement, asked him how he knew it. "Why," [said] he, "I have calculated it," whereupon Dr. Halley asked him for his calculation without any further delay. Sir Isaac looked among his papers but could not find it, but he promised him to renew it [do the calculation over], and then to send it to him.[14]

So Newton went back to work, and in 1687, urged by students and other scientists, he finally published a book describing his discoveries about gravity and motion. Many scholars feel that this book, the *Principia*, or *Mathematical Principles of Natural Philosophy*, is the greatest scientific book of all time. In it Newton showed his mathematical methods step by step and laid out a system that would describe how objects moved anywhere in the universe.

PHILOSOPHIÆ
NATURALIS
PRINCIPIA
MATHEMATICA.

Autore *J S. NEWTON*, *Trin. Coll. Cantab. Soc.* Matheseos Professore *Lucasiano*, & Societatis Regalis Sodali.

IMPRIMATUR.
S. PEPYS, *Reg. Soc.* PRÆSES.
Julii 5. 1686.

LONDINI,

Jussu *Societatis Regiæ* ac Typis *Josephi Streater*. Prostat apud plures Bibliopolas. *Anno* MDCLXXXVII.

Newton's Principia, *published in 1687, outlined his revolutionary discoveries about gravity and motion.*

The Laws of Motion

In addition to the law of universal gravitation, Newton said there were three basic laws of motion. The first, the law of inertia, said that objects will keep doing what they were doing unless some outside force causes their motion to change. That is, an object sitting at rest will remain at rest, while a moving object will keep moving in a straight line. (It is true that if one rolls a ball or spins a bicycle wheel, the motion will stop after a few moments, but that is because of the forces of friction and air resistance. In the vacuum of space, where there is no air, an object will, indeed, keep moving indefinitely until some outside force affects it.)

The second law said that when force is applied to an object, the object will be accelerated, or pushed faster and faster, in the direction of the force. The amount of acceleration is inversely proportional to the mass, or amount of matter in the object. Thus, if someone kicks a soccer ball, which consists of relatively little matter, it might fly a hundred feet, but if the person kicks a bowling ball, which has a relatively great mass, the result is likely to be a motion of only a few inches—and a sore foot.

Finally, Newton's third law said that every action is matched by an equal and opposite reaction. This law of action and reaction explains how a rocket works. The force with which gas is pushed out of the nozzle in the rear of the rocket is matched by an opposite force that pushes the rocket forward.

Newton's three laws formed the basis for what is now called Newtonian mechanics. This system made it possible to look at objects in a physical situation, note their masses, apply a force to the objects, and calculate the resulting motion.

The Nature of Light

Physics does not just study moving bodies. Physicists are also concerned with energy and how it interacts with matter. One form of energy—light—had qualities that particularly interested seventeenth-century scientists. How did light bend when it hit a boundary between two different substances, such as the surface of a body of water? What caused rainbows? Was there only one kind of light, or might there be many different kinds?

By forcing rays of light to pass through a prism, Isaac Newton discovered that "white" light was actually a mixture of many colors. This significant finding enhanced scientists' understanding of the nature of optics.

Isaac Newton was a key contributor to the science of optics, or the behavior of light, as well as to other parts of physics. Most scientists of the time thought that all light was basically white, though other colors might be added to the white as light went through or bounced off different objects. But when Newton passed white light through a prism, a wedge-shaped piece of glass, he found that the light broke up into a rainbow of colors: red, orange, yellow, green, blue, indigo, and violet. "White" light was actually a mixture of colored lights.

As Galileo had done with the results of his motion experiments, Newton put his knowledge of light to practical use. The existing kind of telescope, called a refractor, operated by having a large lens capture light and bend it toward a smaller lens in the telescope eyepiece. Refractors tended to produce images that were smeared with color around the edges. Newton invented a different kind of telescope, called a reflector. In a reflector, incoming light hits a mirror and is then bounced to a lens in the eyepiece. Reflectors made images that were less distorted, for mirrors could be made larger than lenses. Today most big telescopes use Newton's basic design.

Newton and Dutch physicist Christian Huygens disagreed about the nature of light. Huygens believed that "it is inconceivable to doubt that light consists in the motion of some kind of matter."[15] By thinking of light as a series of moving

An Early Description of the Scientific Revolution

Two French scholars, Denis Diderot and Jean Le Rond d'Alembert, wrote an encyclopedia between 1745 and 1772. In it they discuss the influence of Newton and the development of a scientific revolution. This quote is taken from Bernard Cohen's The Newtonian Revolution.

"Newton appeared, and was the first to show what his predecessors had only glimpsed, the art of introducing Mathematics into Physics and of creating—by uniting experiment and calculation—an exact, profound, brilliant, and new science. At least as great for his experiments in optics as for his system of the world, Newton opened on all sides an immense and certain pathway. England took up his views; the Royal Society considered them as their own from the beginning. The academies in France adopted them more slowly and with more difficulty. . . . The light at last prevailed; the generation hostile to these men had died out in the academies and universities. . . . A new generation has arisen; for once a revolution has been started, it is almost always the case that the revolution is brought to fruition [fulfillment] in the next generation."

waves of some very fine substance, Huygens tried to explain how light traveled and how it was bent (refracted) or reflected. Newton, however, thought of light as consisting of streams of tiny particles. He applied laws of motion to the particles and came up with an alternative explanation of light. Through the following centuries each explanation had its defenders and critics. Each theory explained some parts of light's behavior but had trouble with other aspects. In the twentieth century, the theories would be combined.

Newton and the Age of Reason

After the death of Newton, poet Alexander Pope exclaimed, "Nature and Nature's laws lay hid in night / God said, 'Let Newton be,' and all was light."[16] But while Newton was not a particularly modest person, he had summed up his own life by saying, "If I have seen farther than other men, it is by standing on the shoulders of giants."[17] Both statements seem to be true. Newton discovered more fundamental laws of science than anyone had before—or since. On the other hand, Newton was able to do so much because he built his work on the foundation laid by the extraordinary achievements of Galileo and Kepler.

Newton became a symbol of the power of using reason to understand nature. The belief in universal laws of nature quickly took hold. Indeed, during Newton's lifetime the philosopher Benedict de Spinoza declared:

> We feel certain that the forms and qualities of things can best be explained by the principles of mechanics, and that all effects of nature are produced by motion, figure, texture, and the varying combinations of these and that there is no [need for] inexplicable forms and occult [mystical] qualities.[18]

The idea that any area of study might be summed up by a few powerful laws was soon applied to subjects far removed from the physical sciences. For example, Scottish economist Adam Smith, an admirer of Newton, wrote a book called *The Wealth of Nations* in 1776. In it he set forth a few simple laws that he said explained how a nation could build a prosperous economy. In that same year a new nation, the

Isaac Newton exerted extraordinary influence in the scientific revolution. He is credited with discovering more fundamental laws of science than anyone, either before or after his era.

Science and an Industrial Frenzy

Not everyone praised the scientific revolution. English writer Daniel Defoe, author of Robinson Crusoe, *had a mocking view of technological progress in his introduction to* An Essay upon Projects. *Writing in 1697, Defoe describes a time when ambitious businesspeople were starting many technical projects. Defoe's description is quoted in Alan L. Mackay's* Scientific Quotations.

"Necessity . . . has so violently agitated the wits of men at this time, that it seems not at all improper . . . to call it, the Projecting [project making] Age. . . . The Art of War, which I take to be the highest Perfection of Human Knowledge, is a sufficient Proof of what I say, especially in conducting Armies, and in offensive Engines; witness the new ways of Mines [for blowing up enemy positions] . . . , Entrenchments, Attacks, Lodgements [types of fortifications] . . . and a long [list of other] New Inventions. . . . But if I were to search for a Cause, from whence [where] it comes to pass that this Age swarms with such a multitude of Projectors [people who plan projects] more than usual; who besides the Innumerable Conceptions which die in the bringing forth . . . do really every day produce new Contrivances, Engines, and Projects to get Money, never before thought of. . . . [Merchants], prompted by Necessity, rack their Wits for New Contrivances, New Inventions, New Trades, Stocks, Projects, and anything to retrieve the desperate Credit of their Fortunes [make back money they have lost on other ventures]."

United States, was born. Its Declaration of Independence echoed the words of the philosopher John Locke, who had said that the kind of government people should have must be determined by human experience and reason, not the God-given authority of kings. Because of their emphasis on the use of reason and the application of science to many parts of human life, the seventeenth and eighteenth centuries have been called the Age of Reason, or the Enlightenment.

In the nineteenth century, however, a new movement called Romanticism reacted against the claims of science and reason to explain all things. To the first great Romantic, poet William Blake, Newton was still a symbol of power, but it was a lifeless power because it denied feelings. The Romantics did not urge that religion again be made supreme, but they emphasized the free play of imagination and feelings. The debate about whether reason or feeling is more important continues today.

Chapter

3 Striking Sparks

Ancient people had observed two mysterious forces that seemed to make objects move by themselves. One force, which is now called static electricity, caused certain objects to be attracted to one another. For example, if a piece of amber, or fossilized tree sap, was rubbed with a wool cloth, bits of dust or lint would fly through the air and stick to it.

The other force was found in unusual iron-containing rocks called lodestones. According to legend, shepherds in a part of Turkey called Magnesia sometimes found that their iron-tipped staffs stuck to the ground at a spot where a huge lodestone was buried. The force that produced this effect is now called magnetism. The word *magnet* may have come from the name Magnesia.

The ancient Greek astronomer Thales of Miletus observed that when a piece of iron was rubbed against a lodestone, the iron would gain the lodestone's ability to attract other iron objects. He thought that perhaps amber and lodestone had a soul, or life force, within them that could flow into other objects, attracting some and making others able to attract.

Thales of Miletus believed that lodestone, a form of iron ore, possessed a soul that would attract other iron objects. Today, scientists know that this attraction is based on magnetism.

Although they still did not know how magnetism worked, European navigators were using magnetic compasses by the thirteenth century to help them sail their ships in the open ocean out of sight of land. They may have learned about this invention from the Arabs. A compass was helpful in navigation because its needle always pointed north.

Gilbert's *Terrella*

In the late 1500s an English doctor named William Gilbert became fascinated by magnetism. In particular, he wanted to know why compasses worked the way they did. He began his experiments by making a lodestone sphere that was called a *terrella,* or "little earth." He observed:

> In the heavens the astronomers give to each moving sphere two poles; thus do we find two natural poles of exceeding importance even in our terrestrial [earthly] globe. . . . In like manner the lodestone has from nature its two poles, a northern and a southern; fixed, definite points in the stone, which are the primary [terminals] of the [stone's] movements and effects and the limits and regulators of its several actions and properties [qualities].[19]

Gilbert marked the north and south magnetic poles of the *terrella* and then drew a series of lines between them, like meridians on a globe. By moving a free-hanging needle along one of the lines, he showed that the needle stood in a horizontal position at what would be the equator of the *terrella.* As he moved the needle toward one of the poles, it began to turn. At either pole, the needle was vertical, pointing straight at the pole.

Gilbert's most important conclusion was that the compass worked because the earth itself was a giant magnet. Like the *terrella,* he said, the earth had a north and a south magnetic pole. Invisible lines of magnetic force moved the needle of the compass so that it always pointed toward a pole.

Gilbert also investigated the attraction of an electrically charged piece of amber

After conducting experiments with his terrella, *a lodestone model of the earth, William Gilbert determined that the earth was a giant magnet.*

for nearby objects, which seemed similar to the attraction of a magnet for pieces of iron. To aid his experiments, Gilbert built an instrument that he called a *versorium,* from a Latin word meaning "to turn." The *versorium* was a small bar of material that would turn when attracted by amber. By measuring the amount of turn, Gilbert could compare the amount of force exerted by amber and other substances that shared amber's properties.

Gilbert made a list of materials that proved to have an attractive force when rubbed. The list included glass, sulfur, wax, diamond, and arsenic. Gilbert called such materials electrics, from the Greek word for amber, *elektron*. He called the pull *vis electrica* or "electrical force." Thus our word *electricity* was born.

Gilbert had made an important contribution to the development of the scientific method. His *terrella* was an example of a model—something that could be used in the laboratory to represent natural forces. Furthermore, Gilbert's *versorium* was one of the first instruments deliberately made for scientific use. It showed that scientists not only could learn more about nature but could also invent tools to further their knowledge. Making models is still important in science, but today the models are often computer programs rather than physical objects.

Electricity in a Bottle

Magnets were easy to experiment with because they pulled steadily, but the static electricity from a material like amber worked only in brief bursts. Around 1650, however, a German physicist named Otto von Guericke connected a sulfur-coated ball to a crank. When he turned the crank and held his hand against the ball, static electricity was generated, along with

William Gilbert demonstrates his experiments on electricity to the queen of England and her court.

flashes of light. Soon more elaborate and powerful electricity-generating machines were built. If cranked vigorously, they could generate enough static electricity for experiments. Unlike the electrical current that flows in houses today, however, static electricity could not be used for practical purposes. It came in spurts that were too brief to provide a useful light or turn a motor, for example.

The actual nature of electricity remained a mystery. Because it could flow from the generating machine along a piece of wire, most experimenters thought of electricity as a kind of invisible fluid. In 1745, therefore, E. G. von Kleist, the bishop of Pomerania, now a part of Germany, decided to see whether electricity could be collected and stored in a bottle the way a fluid could.

Von Kleist led a wire into a bottle of water through a cork, then charged the water with an electrical machine. He disconnected the machine from the jar. Later, when von Kleist touched the wire, it discharged the stored electricity, and he received a powerful shock that he later reported, "stunned his arms and shoulders."[20]

This electrical bottle was called the Leyden jar, after the Dutch town where other experimenters had created something similar. Experimenters could connect a Leyden jar to a machine that generated static electricity, crank the machine to produce a charge in the water inside the jar, and store the charge in the bottle for later use. By combining a group of charged Leyden jars, they could start fires or even melt wire. The Leyden jar also became popular for a new party game: A circle of people held hands while one of them held and discharged a jar. The shock, startling but harmless, instantly passed around the circle from hand to hand.

Franklin Tames Lightning

As scientists learned more about static electricity, some began to wonder whether the awesome and frightening bolts of lightning from a thunderstorm might have some connection with the sparks in their Leyden jars. The answer came not from the scientists of the established countries of Europe, but from a remarkable man who lived in the English colonies in America.

In 1746 an American printer named Benjamin Franklin attended a demonstration of electrical wonders in Boston. He was so fascinated that he bought the lecturer's equipment and set up his own laboratory. Franklin found that while a number of interesting experiments had been done with electricity, no good theory explained how electrical charges behaved.

A basic question about electricity was why there seemed to be two kinds of electric charge. If one rubbed a piece of amber with fur and then rubbed a piece of glass with silk, the amber and glass would attract each other. But two pieces of amber (or two of glass) treated in this way would repel each other. Charles du Fay in Paris had proposed that there were two kinds of electrical fluid, which he called vitreous (glassy) and resinous (similar to amber in being a kind of resin or sap).

Franklin came up with a simpler explanation that required only one kind of electrical fluid. He said that some substances,

Benjamin Franklin revolutionized the study of electricity when he theorized that both positive and negative electrical charges exist.

such as glass, gained electrical fluid when rubbed, making them positively charged. Other bodies, such as amber, lost fluid and became negatively charged. Franklin used the mathematical plus and minus signs to stand for positive and negative charges. Franklin's theory proved to be basically correct, and his way of speaking about electricity still can be seen in the + and - symbols on batteries.

Franklin then looked more closely at what happened when something became electrically charged. He found that when a charged object that ended in a point was connected to the ground with a conductor of electricity such as an iron rod, electricity from the atmosphere would be drawn to the point and carried down into the earth by the conductor.

Franklin decided that if he set a pointed rod in a high place and connected it to the ground, the rod should be able to draw or attract any electricity in the atmosphere, even lightning itself. In 1752 Franklin set up a rod in his house in Philadelphia "to draw the lightning down into my house, in order to make some experiments on it, with two bells to give notice when the rod should be electrify'd."[21] Franklin wrote about his lightning research in his *Poor Richard's Almanac* in 1753. He explained that he had shown that lightning behaved in the same way as the electricity that experimenters had made with machines. Lightning, therefore, must be a form of electricity.

The Kite Experiment

A rod, even on a tall building, was limited to a height of a hundred feet or so. To find out whether the atmosphere above this level contained electricity, Franklin flew a kite in a Philadelphia park in 1752 during a thunderstorm. Franklin and his son stood in the shelter of a shed and held the end of the kite string, which was tied to a wet hemp cord, the conductor, and separated from them by a piece of silk, an insulator. A key dangled from the connection between the cord and the silk. When Franklin brought his hand near the key, sparks flew to his knuckles. Soon other experimenters were using kites and rods to bring static electricity down from the sky—sometimes with unfortunate results, because lightning can give a fatal electric shock.

Now that Franklin had shown that lightning could be conducted, he realized that a rod could be used to draw lightning safely away from a building and into the

Benjamin Franklin and his son take shelter in a shed while performing the now famous kite experiment. From this endeavor, Franklin discovered that lightning could be successfully conducted.

ground. In 1760 he set up a demonstration lightning rod in Philadelphia. It was promptly hit by lightning, and both the rod and the building on which it stood survived the lightning strike without harm. Some religious people felt that lightning expressed God's displeasure and thought that attempting to shunt it aside defied God. Gradually, however, as people saw that buildings equipped with lightning rods survived storms, the new device came into widespread use.

Franklin's work is important in the history of science because it showed how theory and experiment could lead to practical technology. Franklin studied static electricity, developed a new way to think of it, and built special equipment to perform new experiments on it. He then used the results of his experiments to design the lifesaving lightning rod.

Electricity on Demand

Franklin and other scientists of his time had to work with static electricity because it was the only type of electricity that was then known. Around 1791, however, an Italian scientist named Alessandro Volta discovered another form of electricity that was easier to study and control.

Volta and other experimenters had occasionally observed that electricity started to flow when two different kinds of metal were accidentally brought near each other. Volta decided to explore this phenomenon, or unusual occurrence, in more detail. He found that he could stack plates made of two kinds of metal in alternating layers, making a sort of sandwich,

In about 1791, Alessandro Volta invented the battery, a device which produced a steady flow of electricity.

and soak them in a solution of salt or acid. Such liquids can conduct electricity. In this voltaic pile, electrons—the fundamental particles of electricity—moved from one piece of metal to the other, carried by the conducting solution. The result was a steady flow, or current, of electricity. A wire attached to one of the pieces of metal could carry the electricity out of the pile to another device, such as a bell.

Volta had invented the first battery. With batteries, scientists could experiment with steadily moving electricity—a current flowing in a circuit—instead of static or motionless electricity—a charge that stays in one place until it jumps from one object to another. Further, the moving electricity from batteries would provide a reliable source of power for the electrical inventions of the next century.

In 1795, shortly after Volta's discovery, a writer named Tiberius Cavallo published a three-volume history of electricity. Looking toward the new century, he remarked:

> Since the time of this discovery [of the Leyden jar], the prodigious [great] number of Electricians, experimenters and new facts that have been daily produced from every corner of Europe, and other parts of the world, is incredible. . . . The young Electrician has a vast field before him, highly deserving his attention, and promising further

Fun with Electricity

In a letter to a friend named Peter Collinson written in 1759, Benjamin Franklin describes an imaginative "electrical party" he is planning. Franklin's letter is quoted in A Treasury of World Science *by Dagobert D. Runes.*

"Chagrined [dismayed] a little that we have been hitherto [until now] able to produce nothing in this way of use to mankind and the hot weather coming on, when electrical experiments are not so agreeable, it is proposed to put an end to them for this season, somewhat humorously, in a party of pleasure on the banks of Skuylkil [River]. Spirits [alcohol], at the same time, are to be fired by a spark sent from side to side through the river, without any other conductor than the water; an experiment which we have some time since [ago] performed, to the amazement of many.

A turkey is to be killed for our dinner by the electrical shock, and roasted by the electrical jack [spit], before a fire kindled by the electrified bottle; when the health of all the famous electricians in England, Holland, France, and Germany are to be drank in electrified bumpers [mugs], under the discharge of guns from the electrical battery."

discoveries, perhaps equally or more important than those already made.[22]

Cavallo's prophecy would soon be fulfilled.

The Secret of Electromagnetism

In 1820 a Danish college professor's demonstration with a battery completely changed the way scientists thought about both electricity and magnetism. Hans Christian Oersted simply wanted to show his university class in Copenhagen how the current from a battery could heat up a wire. A compass, however, happened to be sitting near the wire Oersted was using. To the professor's astonishment, the compass needle began to swing back and forth as soon as he connected the battery to the wire. After a moment the compass needle was pointing straight at the electrified wire. Oersted realized that the moving electric current must be generating a magnetic field that moved the compass needle. This meant that electricity could somehow produce, or be turned into, magnetism.

In London about ten years later, a British scientist named Michael Faraday carried Oersted's insight a step further. If electricity could produce magnetism, Faraday reasoned, it should also be possible to convert magnetism to electricity. Finding a way to do this proved to be anything but easy, however. Faraday and other experimenters tried putting magnets near wires, but nothing seemed to happen. Finally, in 1831, Faraday demonstrated that a *turning* magnet would induce an electric current, or cause a current to flow, in a nearby wire. In other words, a magnet had to be moving in order to produce an electric current.

In 1820 the Danish professor Hans Christian Oersted discovered that electricity generated magnetism.

Faraday's and Oersted's discoveries showed that electricity and magnetism were, in a sense, mirror images of each other. Building on these discoveries, later scientists proved that electricity and magnetism are actually two aspects of the same force, called the electromagnetic force.

Faraday's work also had practical value. His device with the turning magnet was the ancestor of the generators used in today's power plants. In a generator a source of power, such as a steam engine, is used to turn a magnet near a coil of wire. The electricity induced in the wire can then be carried by other wires to wherever people want to use it.

In the course of trying to improve the generator, Faraday found out that the largest amount of current was produced when the moving magnetic field, which consisted of invisible curved lines of force, cut across the wire at a right angle. Using a powerful source of magnetism and making the wire into a tightly wound coil also greatly increased the amount of power produced.

A second invention of Faraday's, which drew on Oersted's discovery that electricity could generate magnetism, led to the electric motors used in so many ways today. Faraday's first motor was simply a piece of wire that could turn freely. The wire could be made to spin as the electricity in a nearby coil of wire induced a magnetic field. Such a device is called an electromagnet. The magnetism pulled the floating wire until it, like Oersted's compass, was aligned with the magnetic field. At this point a switch caused the electrical current, and thus the magnetic field, to change direction, which kept the wire spinning as it tried to realign itself. During the nineteenth century the electric motor, like the electric generator, would

A Dazzling Tower of Lights

A young woman named Mabel Barnes describes the displays of electric lighting that she saw while visiting the Pan-American Exposition of 1901. Her impressions are quoted in David Nye's Electrifying America.

"Through some wonderful mechanism . . . the light comes on by degrees, and this creates a novel effect. The time fixed for the ceremony of illumination is half past eight just as the summer twilight is deepening into darkness. A few minutes before the appointed hour, the bulk of electrical lighting along the paths and within the buildings diminishes until they become tiny specks of flames which soon die away. Suddenly the buildings, the long lines of lamp pillars, seem to pulse with a thrill of life before the eye becomes sensibile [receptive] to what is taking place. There is a deep silence and all eyes are intent on the Electric Tower. In the splendid vertical panel there is a faint glow of light, like the first flush which a church spire catches from the dawn. This deepens from pink to red, and then grows into a luminous yellow, and the Exposition of beams and staff [the appearance of the tower's structure] has vanished and in its place is a wondrous vision of dazzling wonders and minarets, domes and pinnacles set in the midst of scintillating gardens—the triumph not of Aladdin's lamp, but of the masters of modern science over the nature-god Electricity."

be improved as experimenters learned the best way to wind the wire coil.

Electricity Changes the World

The scientific revolution in electricity came much later than that in astronomy or physics. When it finally arrived early in the nineteenth century, another revolution—an industrial one—was in full swing. The two revolutions soon joined forces.

Electric power, unlike power from unpredictable steam made from smoky coal, was clean and easy to control. Once the electric motor was made efficient and reliable, it became a blessing for industry and transportation. By the late nineteenth century electric-powered streetcars were linking growing cities in America and in Europe. Electric generators eventually made it possible for people's homes to receive electricity, too. Thomas Edison's electric light, which was safer and brighter than gaslight, gave people a good reason to want electric power.

Samuel Morse displays his telegraph, an invention that united the modern world through instant communication.

Scientist Michael Faraday expanded on the findings of Oersted and determined that a rotating magnet produced an electric current. Faraday's discovery led to the invention of the electric generator and the electric motor.

Only a generation after Oersted first demonstrated the connection between electricity and magnetism, an inventor named Samuel Morse figured out how to turn the on-off switching of a current into a signal that could pull a magnet. The result was the telegraph, the beginning of the instant communications that tie the modern world together. Indeed, electricity became the very symbol of the modern world. As science was swiftly turned into new technology, it told people not only that knowledge was power but that knowledge harnessed power.

Chapter

4 Nature's Building Blocks

It was a mysterious substance. It was everywhere, yet scientists could not see it, feel it, or taste it. For a long time they did not fully realize what it was. But their attempts to understand it helped to bring the scientific revolution to chemistry.

The mystery material was air.

People had always known air was important, of course. Air was one of the four elements, or fundamental materials, that Aristotle said made up the world: earth, air, water, and fire. A hundred years after Copernicus had banished ancient Greek ideas from astronomy, most chemists still believed in Aristotle's four elements.

Scientists who studied the nature of matter and the way substances combine with and change each other did not call themselves chemists in those days. Well into the seventeenth century they were known as alchemists. Alchemy, like astrology, was science mixed with large doses of religion and magic. Among other things, alchemists thought one element could be changed into another. Some alchemists tried to find a way to change lead into gold. Others hunted for a medicine that would cure all illnesses.

Artisans in early industries had practical knowledge of certain chemical reactions. They knew how to combine, or alloy, metals, how to make dyes from plants, and so on. But although they knew *what* worked, they did not know *why*. There were so many reactions, all seemingly so different, that it was much harder to find a system that explained them than it was to find one that explained physics or astronomy. Thinkers devised many systems, but they seldom tested them by experiments. They could not agree on such basic questions as what matter was made of or why chemical reactions took place.

A Skeptical Chemist

One of the first scientists to try to take the magic out of chemistry was Robert Boyle, the son of a wealthy Irish earl. Boyle was a founding member of the Royal Society, England's great scientific society.

Boyle knew more about air than most of his contemporaries. He became famous for his experiments on air pressure, which he called "the spring of the air." He also showed that air was necessary for both burning (combustion) and breathing (respiration) of mammals. When Boyle placed a burning candle in a sealed jar, the candle soon went out. When he put a mouse in a similar jar, the mouse died if the air in the jar was removed by an air pump. It

Robert Boyle is often called the father of chemistry. Through observation and experimentation, Boyle brought chemistry out of the realm of magic and religion and made it a respectable science.

also died simply after remaining in the jar for a while, as if its breathing had used up the air or something in the air that was necessary for life.

Partly as a result of his experiments with air, Boyle came to believe that all matter was made of extremely tiny particles. He called them corpuscles, which means "small bodies." His ideas were similar to those of the ancient Greek thinker Democritus, who had written that all matter was made of invisible objects that he named atoms. The word *atom* comes from a Greek word meaning "that which cannot be cut" or divided. Democritus's ideas had long been out of fashion, but Boyle helped to revive interest in them.

Boyle published his ideas in a book called *The Skeptical Chymist* in 1661, the same year that the Royal Society was founded. In this book he showed the flaws in the ideas of Aristotle and the alchemists. For example, he pointed out that some substances could be broken down into more than four indivisible products. This suggested that the four-element theory was wrong. Boyle himself defined elements as "certain primitive and simple, or perfectly unmingled bodies"[23] out of which all other substances were made. He said that chemical compounds—substances that contained more than one element—were made up of clusters of corpuscles of different elements. He did not think it was possible to actually isolate an element, however.

Boyle tried to turn chemists' thinking away from magic and into the practical form, based on observation and experiment, that had already revolutionized astronomy and physics. He wrote that he wanted to create "a good understanding betwixt [between] the chymists and the mechanical philosophers"[24] such as Isaac Newton, and for the most part he succeeded. Because of this, Boyle is often called the father of chemistry.

The Phlogiston Theory

Robert Boyle did a good job of showing what was wrong with the ideas of Aristotle and the alchemists, but he did not have a convincing system to put in their place. For the next century, therefore, many chemists still clung either to the older chemical theories or to a new one proposed by a German chemist, Georg Stahl, in 1718. Stahl stated that all substances that could burn contained an element called phlogiston (from a Greek word meaning something that can be burned). Phlogiston was believed to be released into the air during burning. The phlogiston theory seemed to explain not only burning but many other kinds of chemical reactions. But it was wrong.

The man who put an end to the phlogiston theory was Antoine Laurent Lavoisier, a Frenchman born in 1743. In the nationalistic nineteenth century, when each European country tried to claim a chief place in science, a scientist writing in France went so far as to state, "Chemistry is a French science; it was founded by Lavoisier of immortal fame."[25] He exaggerated, but perhaps not by much.

Like Boyle, Lavoisier was rich. Part of Lavoisier's wealth was inherited, and part came from his job in the Ferme Général, a private company that collected taxes for the French government. Lavoisier also helped the government by inventing a more efficient way to make gunpowder and by making weights and measures more accurate.

French scientist Antoine Laurent Lavoisier was influential in the establishment of modern chemistry, especially the study of elements and gases.

Most of Lavoisier's time and money, however, went into science. He built his own laboratory, complete with complex and expensive instruments. He exchanged letters with scientists in other countries, such as Benjamin Franklin. He was made a member of the French Academy of Sciences, an organization of the country's top scientists, when he was only twenty-four years old.

Added or Taken Away?

In 1772 Lavoisier became curious about what happened to metals and to the air around them when the metals were heated. The phlogiston theory said that the metals released phlogiston into the air, leaving behind a powdery substance. To test this idea, Lavoisier did an experiment with mercury, a silvery metal that is a liquid at room temperature. (Liquid mercury is the silvery substance in most thermometers.)

In his experiment Lavoisier put a carefully weighed quantity of mercury in a vessel whose only opening joined a jar that contained a measured volume of air. The jar was closed at the top and open at the bottom. The bottom was placed in a dish of mercury and thus sealed off from the atmosphere.

Lavoisier then heated the vessel containing the weighed mercury. The silvery liquid slowly changed to a reddish powder. At the same time, the level of the mercury

Lavoisier conducts an experiment in his laboratory. During his career, Lavoisier made great strides in the field of chemistry by disproving the phlogiston theory.

in the jar rose. It was drawn up from the dish into the jar because the volume of air in the jar had decreased. By the time the changes stopped, one-fifth of the original volume of air in the jar had seemingly vanished.

This experiment suggested to Lavoisier that the phlogiston theory was almost exactly backwards. Instead of something being added to the air during heating or burning, as the theory claimed, something seemed to be taken out of the air and added to the material being burned. But what was that something?

Elements and Gases

Lavoisier found the answer to his question when a British chemist, Joseph Priestley, visited him in 1774. Priestley told Lavoisier about his recent discovery of what he called a "kind of air" in which a candle burned more brightly and quickly than it did in ordinary air. Priestley, who believed in the phlogiston theory, called the new gas dephlogisticated air—air from which all phlogiston had been removed. He thought the candle restored phlogiston to the air when it burned.

Lavoisier, however, suspected that Priestley's gas was what was taken out of the air and added to substances when they burned. He gave it a new name, oxygen, which means "acid former." Acids are sour-tasting substances that are one of the major classes of chemical compounds. Lavoisier believed, incorrectly, that oxygen was a part of all acids.

Lavoisier described his discoveries about burning and much more in a book called *Elements of Chemistry,* which appeared in 1789. In this book he defined a chemical element as "the last point which analysis is capable of reaching"[26]—that is, a substance that cannot be broken down into others by any known means. This sounds similar to Boyle's definition, but unlike Boyle, Lavoisier had specific chemicals in mind. His book listed thirty-three elements, including light and heat. Of these, twenty-three are still considered elements.

Lavoisier's book also presented several important facts about air and other gases. Some of these had been discovered by other chemists, but Lavoisier brought them together and helped to popularize them. First, he stated that gases were a basic form, or state, of matter, like solids and liquids. Solids and liquids could be changed into gases by heating. He pointed out that air was a mixture of different gases, not a single gas. And, finally, he showed that gases, including those in air, could take part in chemical reactions. Chemists before Lavoisier's time had been unable to describe reactions accurately because they had not recognized the role gases played in them. It was as if they had been trying to solve a puzzle with a third of the pieces missing.

Chemistry by Measurement

"The usefulness and accuracy of Chemistry depend entirely upon the determination of the weights of the ingredients and products"[27] of reactions, Lavoisier wrote. For this task he used a chemical balance, an instrument that compares the weight of a substance on one side of it with known weights on the other side. One of Lavoisier's balances was so accurate that it could measure one one-hundredth of the weight of a drop of water.

Weighing chemicals was important, Lavoisier believed, because matter is neither created nor destroyed during chemical reactions. He wrote:

> It can be taken as an axiom [unquestionable fact] that in every operation [chemical reaction] an equal quantity of matter exists both before and after the operation, that the quality and quantity of the principles [elements] remain the same and that only changes and modifications occur. The whole art of making experiments in chemistry is founded on this principle.[28]

As Lavoisier's mercury experiment had shown, measurement helped chemists learn exactly what went on during reactions.

Three States of Matter

Antoine Lavoisier's description of the three states of matter—solids, liquids, and gases—is quoted in The Norton History of Chemistry *by William H. Brock.*

"All bodies in nature present themselves to us in three different states. Some are solid like stones, earths, salts, and metals. Others are fluid like water, mercury, spirits of wine; and others finally are in a third state which I shall call the state of expansion or of vapours, such as water when one heats it above the boiling point. The same body can pass successively through each of these states, and in order to make this phenomenon occur it is necessary only to combine it with a greater or lesser quantity of the matter of fire."

A Revolution in Language

Finally, Lavoisier produced what he called "a revolution of physics and chemistry"[29] by changing the language of chemistry. Before his time, chemists had named compounds with descriptive terms such as *flowers of sulfur*. Such terms told little about what elements the compounds contained. Sometimes the same compound had several names.

Lavoisier proposed that all compounds be given names that showed what was known about their composition. The names would consist of the known elements, plus certain prefixes and suffixes. For example, a compound name made up of the names of two elements plus the suffix *-ide* shows that only two elements are present in the compound. Hydrogen sulfide is a compound made of hydrogen and sulfur; mercuric oxide contains only mercury and oxygen.

Other chemists quickly adopted Lavoisier's system because the names it produced were easier to remember and less confusing than the old ones. Even more importantly, the new names showed both the nature of compounds and their relationship to one another. They allowed chemists to make good guesses about how to make compounds or what a reaction between certain compounds might produce. And when the chemists adopted Lavoisier's naming system, they also automatically adopted his way of thinking about chemistry.

Unfortunately, Lavoisier's revolution in chemistry was cut short by another revolution. In 1789 the French people revolted against King Louis XVI and his corrupt, extravagant court. The revolution began with legitimate grievances, but it

Language and Science

In Elements of Chemistry, *Antoine Lavoisier explained the importance of the language used to describe scientific facts. Lavoisier's explanation, translated by Robert Kerr, is quoted in William Brock's* Norton History of Chemistry.

"Every branch of physical science must consist of three things, the series of facts which are the objects of science, the ideas that represent these facts, and the words by which these ideas are expressed. Like three impressions of the same seal, the word ought to reproduce the idea and the idea to be a picture of the fact and as ideas are presented and communicated by means of words, it necessarily follows that we cannot improve the language of any science without at the same time improving the science itself. Neither can we, on the other hand, improve the science without improving the language or the nomenclature [system of naming] which belongs to it."

turned into a bloody reign of terror in which guilty and innocent alike were executed. Lavoisier was arrested along with the other members of the Ferme Général and was beheaded on May 8, 1794. On hearing of his death, the French mathematician Joseph Lagrange said, "It required only a moment to sever his head, and probably a hundred years will not suffice [be enough] to produce another like it."[30] Luckily, Lavoisier's ideas did not die with him.

John Dalton, the renowned inventor of the modern atomic theory.

Mixtures of Air

A generation after Lavoisier, another man made chemists' understanding of matter even more precise. He was John Dalton, a British schoolmaster who had begun teaching when he was only twelve years old.

Like Boyle and Lavoisier, Dalton made his discoveries partly because of his interest in air. Dalton began studying air because he was fascinated by weather. He carefully recorded the weather every day of his adult life, eventually making over two hundred thousand observations.

By Dalton's time, scientists knew that air was composed of nitrogen, oxygen, carbon dioxide, and water vapor. Dalton wondered why these gases remained mixed together, rather than separating into layers of different weights. He eventually decided, as Boyle and some other scientists had, that all chemicals, including the different gases in air, were made up of extremely small, indivisible particles. Isaac Newton had proposed that the particles of a single gas repel, or push away, each other. But Dalton concluded, "The particles of one gas are not repulsive [repellent] to the particles of another gas, but only to the particles of their own kind."[31] The gases therefore could remain mixed.

From Air to Atoms

In a lecture given in 1809, Dalton described how his thoughts about air led him to begin thinking about the building blocks of matter.

> It occurred to me that I had never contemplated the effect of *difference of size* in the particles of elastic fluids [gases]. . . . This idea occurred to me in 1803. . . . The different *sizes* of the particles of elastic fluids under like [similar] circumstances of temperature

> and pressure being once established, it became an object [purpose] to determine the relative *sizes* and *weights*, together with the relative *numbers* of atoms in a given volume. This led the way to [determining] the combinations of gases and to the number of atoms entering into such combinations.[32]

Dalton described his ideas about matter in *A New System of Chemical Philosophy*, of which part one appeared in 1808. All matter, he said, is composed of extremely tiny particles that cannot be divided. Dalton called these particles atoms, borrowing Democritus's term. The atoms of each chemical element, Dalton said, are all alike, but they are different from the atoms of other elements. Atoms of different elements can combine with each other to form clusters that Dalton called compound atoms. Later chemists called them molecules. A molecule is the smallest unit of a compound. Atoms cannot be created, destroyed, or changed during chemical reactions. Only their combinations are changed. Dalton's name came to be so identified with this atomic theory that one writer in the nineteenth century jokingly defined atoms as "round bits of wood invented by Mr. Dalton."[33]

How to Weigh an Atom

Unlike Democritus and Boyle, Dalton saw his atoms as real, physical things. Most importantly, he said that they had mass, or weight. All atoms of the same element had the same weight, but the weights of the atoms of different elements were different, he believed. Dalton even figured out a way of measuring atomic weights. He could not actually weigh an atom, but he found that he could assign weights relative to each other.

Suppose, Dalton said, that elements X and Y can form two compounds. A molecule of one compound contains one atom of X and one atom of Y. The chemical formula of that compound is XY. A molecule of the other compound contains two atoms of X and one of Y. This compound's formula is therefore X_2Y. If the two compounds were broken down into X and Y, Dalton reasoned, there would be twice as great a weight of X per gram of Y in the second compound as in the first. If a compound's formula were X_3Y, the compound would contain three times as much X per gram of Y as XY. In short, the ratios of elements in a compound should be small whole numbers.

Dalton tested his theory with gases. He knew about three compound gases that contained only nitrogen and oxygen. When he measured the weight of oxygen per gram of nitrogen in each compound, he found ratios of 1, 2, and 4 to 1. For two gases made up of oxygen and carbon, he found that the weights of oxygen per gram of carbon were in the ratios of 1 and 2 to 1. Other tests produced similar results.

Once Dalton had identified compounds that seemed to contain only one atom of each element per molecule, he could conclude that the ratio of the weight of one element to the weight of the other in those compounds was the same as the ratio of their atomic weights. He gave hydrogen, the lightest substance known, the arbitrary atomic weight of one. He made all other atomic weights multiples of this.

Some of Dalton's assumptions about chemical formulas were wrong, and some of his measurements were in error. As a result, his atomic weights differ from those used today. But the idea of making a table of such weights was very important. It allowed chemists to predict how much of certain chemicals would be needed to make compounds with other chemicals. It meant they could write equations for reactions that are as precise as the equations in algebra.

Classifying the Elements

The final step in making chemistry an orderly science came about sixty years after Dalton's book, when Dmitry Mendeleyev, a Russian scientist, arranged the chemical elements—by then there were sixty-three known ones—in a new kind of table according to atomic weight. Mendeleyev had

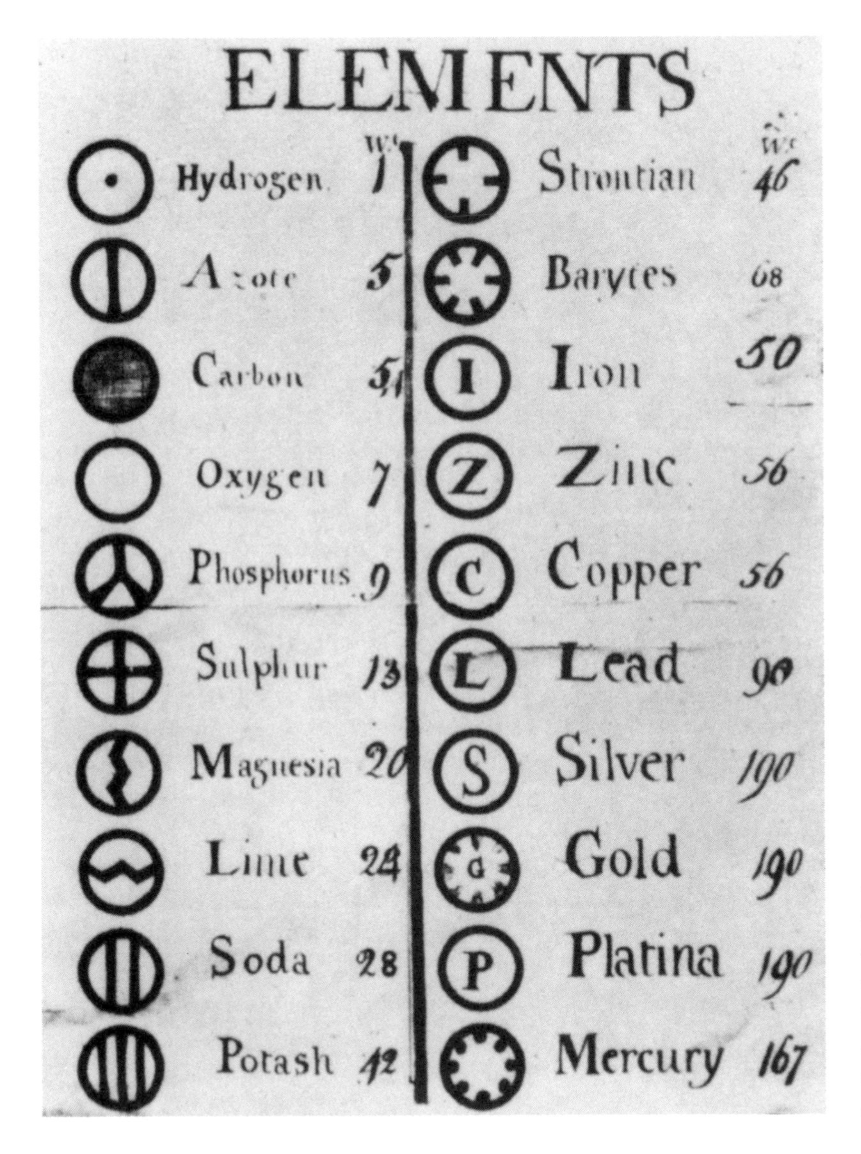

ELEMENTS

	Wt		Wt
Hydrogen	1	Strontian	46
Azote	5	Barytes	68
Carbon	54	Iron	50
Oxygen	7	Zinc	56
Phosphorus	9	Copper	56
Sulphur	13	Lead	90
Magnesia	20	Silver	190
Lime	24	Gold	190
Soda	28	Platina	190
Potash	42	Mercury	167

After determining relative atomic weights, chemist John Dalton compiled the first table of elements (pictured). This table helped chemists ascertain the amount of elements needed to make different compounds.

Russian scientist Dmitry Mendeleyev formulated the Period Law, which stated that the properties of elements vary periodically when arranged by increasing atomic weight.

noticed that elements close together in atomic weight often had many qualities in common. "All the comparisons which I have made . . . lead me to conclude that *the size of the atomic weight determines the nature of the elements,*"[34] he wrote.

Mendeleyev laid out his table in seven rows. The elements in each row shared certain qualities, or properties. Most rows contained either seven or eight elements. Because some other qualities repeated themselves at regular intervals, or periodically—usually after every seven elements — Mendeleyev's table, first published in 1869, came to be called the periodic table.

The most creative thing about the periodic table was that Mendeleyev left gaps in it. His studies of the properties of those elements he knew about led him to predict that at certain points in the table, an as yet unknown element with certain qualities had to exist because the elements on either side of the gap were too different from each other. Mendeleyev's ideas were proved right when the missing elements, such as gallium and scandium, were later discovered. They had exactly the properties he had predicted.

Paths Through a Monstrous Thicket

While Mendeleyev was classifying elements, other chemists were using the knowledge of chemical reactions developed by Lavoisier, Dalton, and others to unravel the puzzles of the most complex part of chemistry: organic chemistry, the study of compounds of the element carbon. Most of these compounds came originally from living things. Organic compounds were made of the same elements as inorganic ones—usually combinations of carbon, hydrogen, oxygen, and sometimes nitrogen. Most organic molecules, however, were much larger and more complex than those of inorganic substances. Until the middle of the nineteenth century, organic chemistry remained almost as mysterious as inorganic chemistry had been in the days of the alchemists.

Chemists were not even sure that the same chemical rules applied to organic and inorganic compounds until 1826, when a German chemist, Friedrich Wöhler, synthesized, or made artificially, the organic substance urea from inorganic compounds. This proved that there was no fundamental difference between the two kinds of compounds. Still, as late as 1885 Wöhler wrote to a fellow scientist:

> Organic chemistry . . . is enough to drive one mad. It gives one the impression of a primeval, tropical forest full of the most remarkable things, a monstrous and boundless thicket, with no way of escape, into which one may well dread to enter.[35]

By that time, however, chemists had hacked some important paths through the monstrous thicket. For example, Friedrich Kekulé von Stradonitz, a German chemist, had discovered in 1858 that each carbon atom could combine with exactly four other atoms. This finding provided a major advance in understanding how the molecules of organic substances were put together. Later Kekulé, as he was usually known, also found out that carbon atoms, often with other atoms attached to them, could form long chains within a single molecule. Sometimes they also formed rings. Kekulé claimed that he first recognized the ring structure of one common organic compound, benzene, because of a dream.

The discoveries of Kekulé and others showed that the three-dimensional arrangement of atoms within molecules was much more important in organic chemistry than in inorganic chemistry. To describe organic compounds properly, chemists had to use formulas that showed not only the kinds and numbers of atoms in each molecule but also the molecule's structure. Indeed, toward the end of the century, several chemists discovered that two organic substances could have molecules with exactly the same numbers and kinds of atoms yet possess entirely different properties because the structure of their molecules was different.

A Rainbow of Synthetics

Once scientists understood the structure of organic molecules, they could begin to make more organic substances artificially. The power to synthesize organic compounds, rather than laboriously extracting them from plant or animal materials, revolutionized several European industries toward the end of the nineteenth century. The dye industry was one of these.

Before the mid–nineteenth century most dyes had come from plants or animals. The ancients had made a deep purple dye from a kind of sea snail, for example. Indigo, a dark blue dye, came mostly from plants grown in India. Because such source materials were limited and extraction was difficult, many dyes were expensive.

All this began to change one day in 1856, when William Henry Perkin, an eighteen-year-old student at the Royal College of Chemistry in London, did a home experiment. Perkin's teacher, August Hoffmann, had told him that several useful organic compounds had been made from coal tar. Coal tar was a black, sticky liquid left after coal was broken down to form coal gas, which was used for lighting at the time. Hoffmann said he suspected that quinine, an important drug that came from the bark of a South American tree, could also be made from coal tar. Perkin decided to find out whether his professor was right.

William Henry Perkin displays a piece of cloth that has been colored with his mauve dye. This synthetic dye became a best-seller and soon replaced the more expensive natural dyes.

The result of Perkin's synthesis—which he started with analine, a compound made from coal tar, rather than coal tar itself—was a blackish mess that bore no resemblance to quinine. When some of the substance spilled and Perkin wiped it up with a cloth, however, he noticed that the cloth turned a lovely shade of pale purple. He showed the cloth to chemists at a large dye factory and asked whether his black compound might be useful as a dye. They thought it might. In time, Perkin's analine purple, or mauve, became a best-selling dye.

Other artificial dyes, most also made from coal tar, soon followed. The first to be made on a scientific basis, using the information about organic chemistry discovered by Kekulé and others, was called alizarin. It appeared in 1869, the same year Mendeleyev published his periodic table. Synthetic dyes quickly replaced many natural dyes because the synthetics were easier to make and, therefore, cheaper.

Other synthetic organic chemicals became varnishes, explosives, and life-saving drugs. These compounds proved very valuable to the companies that made them as well as to the industries and individuals who used them. Business owners began to realize that a knowledge of chemistry could bring high profits. The revolution in chemistry that began with Robert Boyle thus changed not only scientists' basic understanding of matter but also the economy and daily life of nations.

Chapter

5 A Changing Earth

In 1650 an Irish archbishop named James Ussher announced that he knew exactly how old the world was. He used historical documents to figure out when certain events described in the Old Testament of the Bible took place. Then he counted the number of generations that the Bible said had existed between Old Testament times and the time God created human beings, as described in the Book of Genesis. Adding all these generations together, Ussher calculated that God had made the earth and all living things at 9:00 A.M. on October 26, 4004 B.C.

Not everyone agreed with Ussher's precise date. Until the late eighteenth century, however, few Europeans questioned the assumptions he had made in figuring it. Most were sure that the Bible contained a true and literal account of the way the earth began. People also believed that the earth and its plants and animals had not changed in any major way since the great flood described in the Book of Genesis.

A few things made some scientists wonder whether there was more to earth's story than the Bible suggested, however. In some places they could see rocks of different types stacked in layers, like a cake. How had those layers formed? Why were the stacks sometimes tilted almost sideways? And why did they sometimes contain stone objects that looked like seashells? Leonardo da Vinci, for one, thought that these fossils, as they were later called, had once been parts of living creatures. But how could shells have been carried onto mountaintops and turned to stone?

Neptunists and Vulcanists

In the second half of the eighteenth century, a few scientists began to suspect that rocks contained clues to events in the earth's distant past. They did not question the Bible's account of the creation of human beings. They believed, however, that the earth itself was far older than humans—perhaps a million years old. Different groups, studying different kinds of rocks, devised different theories to explain how the earth had changed over time.

One group of scientists, headed by Abraham Werner in Germany, said that an ancient sea had been the main force that shaped the earth. This group came to be called Neptunists, for Neptune, the Roman god of the sea. Another group, centered in France, claimed that eruptions of volcanoes had made or rearranged most of the earth's rocks. They were called Vulcanists, for Vulcan, the Roman god of fire and metalworking. The word *volcano* comes from Vulcan's name.

Neptunists and Vulcanists argued fiercely during the late eighteenth and early nineteenth centuries. By the 1820s, though, most scientists came to agree that there was more evidence for Vulcanism than for Neptunism. As one scientist who converted from Neptunism to Vulcanism put it, "For a long while I was troubled with water on the brain, but light and heat have completely dissipated it [boiled it away]."[36] In fact both seas and volcanoes helped to shape the ancient earth, but the seas did not act in the way Werner described.

Around 1795, while the arguments between Neptunists and Vulcanists were still going on, a Scottish doctor-turned-geologist named James Hutton added some important ideas to Vulcanism. Vulcanists had claimed that the heat of volcanoes came from underground fires that burned coal. Hutton, however, said that the interior of the earth itself was hot enough to melt rock. When the heat built up enough, volcanoes erupted. The molten rock was forced onto the earth's surface.

During the late eighteenth century, geologist James Hutton announced several shocking theories about the earth's formation.

Calculating the Earth's Age

More shockingly, Hutton maintained that the earth was so old that its rocks showed "no vestige [trace] of a beginning—no prospect [likelihood] of an end."[37] Furthermore, he said, no gigantic catastrophes such as the biblical flood needed to be called upon to explain the changes that had taken place in the earth. Forces still in existence, such as volcanoes and earthquakes, were enough.

Many scientists disagreed with Hutton. Georges Cuvier, a Frenchman, was one of the most important. Cuvier startled Parisians around 1800 by showing them that their city was, in effect, built on a cemetery. From rocks around Paris he dug up the fossil bones of elephants, hippopotamuses, and even giant lizards—what we now would call dinosaurs. Honoré de Balzac, a famous French writer, exclaimed:

> Is Cuvier not the greatest poet of our century? Our immortal naturalist has reconstructed worlds from blanched [bleached] bones. He picks up a piece

of gypsum [a kind of rock] and says to us, "See!" Suddenly stone turns into animals, the dead come to life, and another world unrolls before our eyes.[38]

Some of the fossil creatures Cuvier identified were similar to living animals, but others were completely different. In time he described 150 fossil species, of which 90 appeared not to be related to any kind of animal then living. Cuvier believed that the bones he found came from animals that had lived before the flood described in the Bible. He thought God had sent many earlier floods or other catastrophes to earth. Each catastrophe killed most living things, and God created new ones when the disaster was over. "Life has often been disturbed by great and terrible events,"[39] Cuvier wrote.

Cuvier's views were popular, but Hutton's ideas would not die. Another British geologist, Charles Lyell, revived and expanded Hutton's theories in a book called *Principles of Geology*, published in 1830. Like Hutton, Lyell believed that earth had never suffered from giant catastrophes. Instead, he said, all changes seen in the planet's surface could be explained by processes that were still going on:

Georges Cuvier exhibits several fossils to a group of scientists. During the early 1800s, Cuvier discovered 150 fossilized creatures, 90 of which appeared to be extinct.

> No causes whatever have from the earliest time . . . to the present, ever acted, but [except] those *now acting*; and . . . they have never acted with different degrees of energy from that which they now exert.[40]

Just as Vulcanism eventually won out over Neptunism, so Hutton's and Lyell's ideas eventually won out over Cuvier's. Scientists found more and more evidence that seemingly minor changes in the earth's surface, such as erosion by water, could eventually create great effects, such as the carving of gigantic canyons.

If causes "now acting" had been enough to create the huge changes that could be seen in the earth's surface, could they also explain the differences between Cuvier's fossil animals and modern ones? And if they could, what were those causes?

Around 1820 a French scientist named Jean-Baptiste de Monet de Lamarck explained that he thought he knew the answer. Lamarck wrote:

> A number of known facts proves that the continued use of any organ [body part] leads to its development, strengthens it and even enlarges it. . . . Permanent disuse of any organ . . . causes it to deteriorate [break down] and ultimately disappear if the disuse continues . . . through successive generations. Hence we may infer that when some change in the environment leads to a change of habit in some race of animals, the organs that

River of Change

Geologist John Playfair wrote a simplified description of James Hutton's ideas, Illustrations of the Huttonian Theory of the Earth, *in 1802. In this passage quoted in A. Hallam's* Great Geological Controversies, *Playfair shows how forces still in existence can cause major changes in the earth's surface.*

"If . . . a river consisted of a single stream without branches, running in a straight valley, it might be supposed that some great concussion [impact], or some powerful torrent [flood], had opened at once the channel by which its waters are conducted to the ocean; but, when the usual form of a river is considered, the trunk divided into many branches, which rise at a great distance from one another, and these again subdivided into an infinity of smaller ramifications [branches], it becomes strongly impressed upon the mind that all these channels have been cut by the waters themselves; that they have been slowly dug out by the washing and erosion of the land; and that it is by the repeated touches of the same instrument that this curious assemblage of lines has been engraved so deeply on the surface of the globe."

Jean-Baptiste de Monet de Lamarck formulated an early theory of evolution. Lamarck believed that living things could pass characteristics they had acquired during their lifetime to their offspring.

> are less used die away little by little, while those which are more used develop better.[41]

Suppose, for example, a long-term drought made low-growing bushes scarce but had less effect on trees. Grazing animals might stretch their necks to reach the tree leaves that were their only remaining food. As years passed, the necks of the most successful animals would grow longer because of this constant stretching. The animals' offspring would inherit these long necks, Lamarck believed. Thus giraffes might have developed.

A Thought-Provoking Voyage

A late-eighteenth-century English naturalist, Erasmus Darwin, had had ideas similar to Lamarck's. But Darwin's grandson, Charles, proposed a quite different mechanism for the progressive change, or evolution, of living things.

In spite of a poor record in school, as a young man Charles Darwin got a job as a naturalist on a ship called the *Beagle.* In 1831 the *Beagle* began a five-year voyage to survey the coasts of South America and nearby islands for the British government.

Darwin had never felt any interest in his grandfather's views about evolution. His own favorite subject of study was geology, and he even brought a copy of Lyell's *Principles of Geology* with him on the *Beagle* voyage. During his years aboard the *Beagle,* however, he discovered that differences in plants and animals could be just as exciting as those in rocks.

As the *Beagle* sailed down the South American coast, Darwin noticed that the animals he saw in one place were replaced farther along by others that were similar but not quite identical. They belonged to the same genus, or family group, but were of different species, or particular types. Sometimes, instead, they belonged to the same species but differed in some slight way, such as having tails of different length. As Darwin wrote later in his *Autobiography*:

> It was evident that such facts as these could only be explained on the supposition that species gradually became modified [changed slightly], and the subject haunted me. But it was equally evident that neither the action of the surrounding conditions, nor the will of organisms [living things], . . . could account for the innumerable cases in which organisms . . . are beautifully adapted to their habits of life.[42]

A Walk in a Rain Forest

Charles Darwin appreciated the beauty of nature as well as its scientific interest. In this passage from The Voyage of the Beagle, *an account of his South American trip, Darwin describes his feelings while walking through a Brazilian rain forest. The passage is quoted in Benjamin Farrington's* What Darwin Really Said.

"Delight is a weak term to express the feelings of a naturalist who, for the first time, has wandered by himself in a Brazilian forest. The elegance of the grasses, the novelty of the parasitical plants [those feeding off other plants], the beauty of the flowers, the glossy green of the foliage, but above all the general luxuriance [thickness and lushness] of the vegetation, filled me with admiration. A most paradoxical [seemingly contradictory] mixture of sound and silence pervades [exists throughout] the shady parts of the wood. The noise from the insects is so loud that it may be heard even in a vessel anchored several hundred feet from the shore; yet within the recesses [deepest parts] of the forest a universal silence appears to reign. To a person fond of natural history such a day as this brings with it a deeper pleasure than he can ever hope to experience again."

Competition Leads to Selection

Long after he returned to England, Darwin continued to be haunted by the question of how species changed. He read everything he could on the subject. He also learned from breeders how they selected domestic plants and animals with desirable characteristics and mated them to produce offspring with those characteristics. Darwin became convinced that selection was involved in changing wild plants and animals, too. But how did it work? What acted the part of the breeder and made the selection?

The writings of an English clergyman named Thomas Robert Malthus provided Darwin with a clue. In a book written at the end of the eighteenth century, Malthus had maintained that there was so much poverty and misery in the world because the human population grew much faster than its food supply. The same situation, Malthus said, existed everywhere in nature:

> Throughout the animal and vegetable kingdom, nature has scattered the seeds of life abroad [widely] with the most profuse and liberal hand. She has been comparatively sparing in the room and nourishment necessary to rear them. The race of plants and the

> race of animals shrink under this great restrictive law. . . . Its effects are waste of seed, sickness and premature death.[43]

Darwin read Malthus's book in 1836. Having seen the struggle for existence in nature for himself in South America, Darwin wrote:

> It at once struck me that under these circumstances [of struggle for limited food] favourable variations [in plants and animals] would tend to be preserved, and unfavourable ones to be destroyed. The result of this would be the formation of a new species. Here then I had at last got a theory by which to work.[44]

Charles Darwin changed history when he proposed the theory of natural selection, the evolutionary process by which the most adaptable species survive.

Selection Leads to Evolution

Darwin called his theory *natural selection.* It differed from Lamarck's theory in that Darwin did not believe that living things could pass on characteristics acquired during their lifetime. Instead, he thought, changes in species began with chance variations that occurred at birth.

In Darwin's version of the giraffe example, one grazing animal might happen to be born with a longer neck than its brothers and sisters. During a long drought when tree leaves were the main food available, the long-necked animal could get more food than the others. It, therefore, would be more likely to survive and have offspring.

Some of the long-necked animal's offspring probably would inherit its long neck, because offspring tend to be like their parents. If the drought went on, they, too, would get more food and produce more descendants. In time, only the animals that inherited long necks would survive in the drought area. They would become so different from the animals with shorter necks, which might survive in places where the drought was less severe, that the two types could no longer successfully mate with each other. In other words, they would become different species. If the drought was widespread enough, the short-necked species might even die out completely, or become extinct.

The evolution of living things, Darwin believed, happened gradually over long periods of time. "As natural selection acts solely by accumulating slight, successive, favourable variations," he wrote, "it can produce no great or sudden modification."[45] Darwin's view of evolution was thus very similar to Hutton's and Lyell's view of changes in the earth's surface.

A Controversial Book

For twenty years Darwin refined his theory and collected examples to support it. He might have gone on doing so all his life if another English naturalist, Alfred Russel Wallace, had not sent him a paper in 1858 that proposed almost exactly the same ideas. Wallace had worked them out independently. Darwin knew he must now publish his results or let all the credit for the discovery go to Wallace.

Darwin presented a short account of his own discoveries, along with Wallace's paper, to a scientific society on July 1, 1858. Then, in 1859, Darwin published a complete book about his theory. He called it *On the Origin of Species by Means of Natural Selection, or the Preservation of Favoured Races in the Struggle for Life.*

Darwin's book became a best-seller, as scientific books went—and scientific books in the mid–nineteenth century sometimes outsold popular novels. It also aroused a storm of controversy. Some criticism came from religious people, who objected to it because it disagreed with the account of creation in the Book of Genesis. Such people especially disliked the idea that human beings might be descended from animals instead of having been specially created by God. Darwin carefully did not say anything about humans in *Origin of Species*, but the book clearly implied that they had evolved just like other animals.

Other people opposed Darwin's theory on scientific grounds. Some scientists complained that although Darwin provided many examples that supported his ideas, none actually showed one species changing into another. Other writers had questions about the evolution of complex body parts. For example, an arm had to go through many changes as it evolved into a wing. These changes most likely did not all happen at once. But why would natural selection have preserved an arm that was different from other arms but not

The title page from Charles Darwin's The Origin of Species. *Published in 1859, this controversial book detailed Darwin's revolutionary theory of natural selection.*

ON

THE ORIGIN OF SPECIES

BY MEANS OF NATURAL SELECTION,

OR THE

PRESERVATION OF FAVOURED RACES IN THE STRUGGLE FOR LIFE.

BY CHARLES DARWIN, M.A.,

FELLOW OF THE ROYAL, GEOLOGICAL, LINNÆAN, ETC., SOCIETIES;
AUTHOR OF 'JOURNAL OF RESEARCHES DURING H. M. S. BEAGLE'S VOYAGE ROUND THE WORLD.'

LONDON:
JOHN MURRAY, ALBEMARLE STREET.
1859.

The right of Translation is reserved.

This 1871 cartoon mocks Darwin's theory that man evolved from animals. In the caption, the sobbing gorilla laments, "That man wants to claim my pedigree. He says he is one of my descendants."

yet capable of flight? Darwin admitted that he did not have good answers for some of these questions.

From Ridicule to Acceptance

Discussion of both the scientific merits of Darwin's work and its seeming challenge to religion continued for years. But Darwin finally triumphed, just as Copernicus and Galileo had. A professor named William Whewell, who had taught at Cambridge University when Darwin studied there, once said that attitudes toward all great scientific discoveries go through three stages: "It is absurd," "It is contrary to the Bible," and "We always knew it was so."[46] Darwin's theory went through all three stages in the amazingly short time of less than a decade.

In 1871 Darwin published another book, *The Descent of Man*. In this book he explicitly stated that evolution and natural selection applied to human beings. He began by pointing out physical similarities between humans and animals. He then claimed that even people's feelings and mental powers, those things that seem most uniquely human, are not so different from those of animals:

> The great difference in mind between men and the higher animals . . . is certainly one of degree and not of kind. The senses and institutions, the various emotions and faculties [mental

"Grandeur in This View of Life"

Some people felt that Charles Darwin's theory of evolution demeaned God and nature by making change in nature seem merely mechanical. In this passage from Origin of Species, *however, Darwin describes his feeling that a natural world governed by evolution is both complex and beautiful. The passage is quoted in volume 3 of* The History of Science in Western Civilization.

"It is interesting to contemplate an entangled bank, clothed with many plants of many kinds, with birds singing on the bushes, with various insects flitting about, and with worms crawling through the damp earth, and to reflect that these elaborately constructed forms, so different from each other, and dependent on each other in so complex a manner, have all been produced by laws acting around us. These laws, taken in the largest sense, being Growth with Reproduction; Inheritance which is almost implied by reproduction; Variability from the indirect and direct action of the external conditions of life, and from use and disuse; a Ratio of Increase so high as to lead to a Struggle for Life, and as a consequence to Natural Selection, entailing Divergence [difference] of Character [features] and the Extinction of less-improved forms. Thus, from . . . famine and death, the most exalted object [highest purpose] which we are capable of conceiving, namely, the production of the higher animals, directly follows. There is grandeur [magnificence] in this view of life, with its several powers, having been originally breathed into a few forms or into one; and that, whilst this planet has gone cycling on according to the fixed law of gravity, from so simple a beginning endless forms most beautiful and most wonderful have been, and are being, evolved."

abilities], such as love, memory, attention, curiosity, imitation, reason, etc., of which man boasts, may be found in an incipient [undeveloped], or even sometimes in a well-developed condition, in the lower animals.[47]

Oddly enough, *The Descent of Man* produced less uproar than *Origin of Species* had. This was partly because the ideas in it were no longer new and shocking. It may also have been because Darwin's ideas as a whole fit in with the beliefs of the time. Western society in the late nineteenth century glorified competition among individuals, businesses, and nations. Those who succeeded in competition, people believed, were better, or fitter, than those who failed. When Darwin's theory of natural selection was applied to humans, it seemed to elevate these beliefs to the status of a natural law.

Chapter

6 Hidden Worlds

The first modern Europeans to explore the hidden world inside the body risked prison or worse to do it. Cutting open a dead human body was against the laws of the church and of most countries during the Middle Ages. Doctors of this time, therefore, had no good way to learn about anatomy, the structure of the body, except by examining animals.

During the early Renaissance doctors at a few Italian universities gained permission to cut apart, or dissect, the bodies of executed criminals. An assistant carried out the dissection in front of student classes two or three times a year. Meanwhile, the teacher read anatomy descriptions from the writings of Galen, a Greek doctor who had died in A.D. 200. The crowd of students could see little of the dissection. They could not tell—and did not dare ask—whether real bodies matched Galen's descriptions.

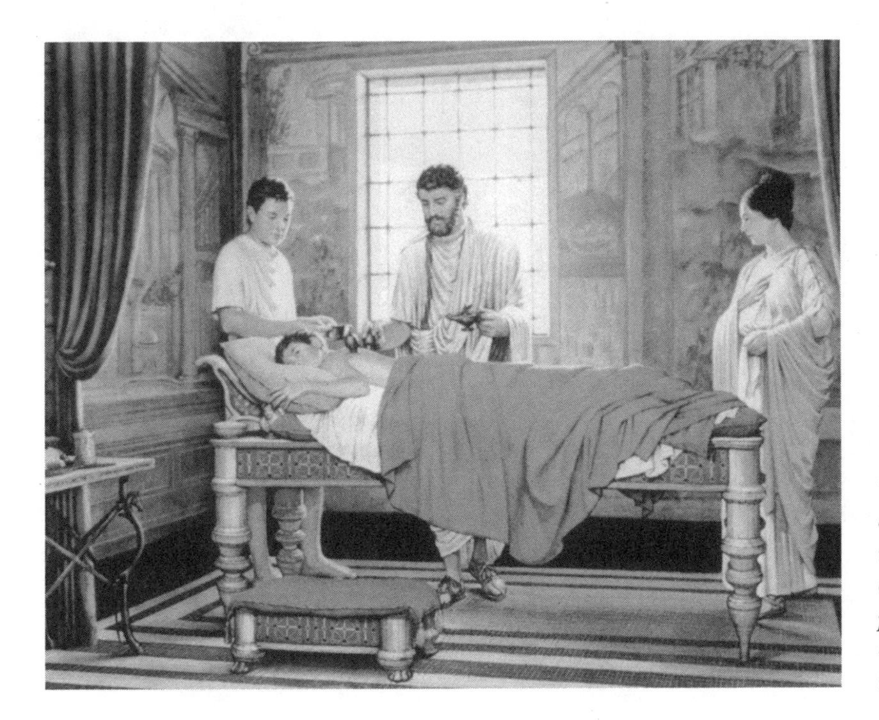

The Greek physician Galen advanced the study of medicine with his observations in anatomy and physiology. He remained an authority on anatomy until the Renaissance.

A New Look at Anatomy

In the early sixteenth century Andreas Vesalius, a Belgian medical student, decided that a few peeks inside pigs or dogs and twice-yearly glimpses of human corpses were not enough. During his university training in Paris, Vesalius stole parts of bodies from the city's public execution spot and dissected them himself. As a deterrent to would-be criminals, corpses of executed people were left hanging in the execution spot until they rotted. Vesalius cut off the parts he wanted and in time assembled a whole skeleton. He became so knowledgeable that he was made a professor of surgery and anatomy at the University of Padua, Italy, when he was only twenty-three years old.

As a teacher in Padua, Vesalius could do dissections legally. He noticed more and more differences between what he saw in human bodies and what Galen had said he should see. The few other doctors who had spotted such differences simply

Andreas Vesalius became an expert in the field of human anatomy. In 1543, he published a revolutionary book that described anatomy and refuted the popular claims of Galen.

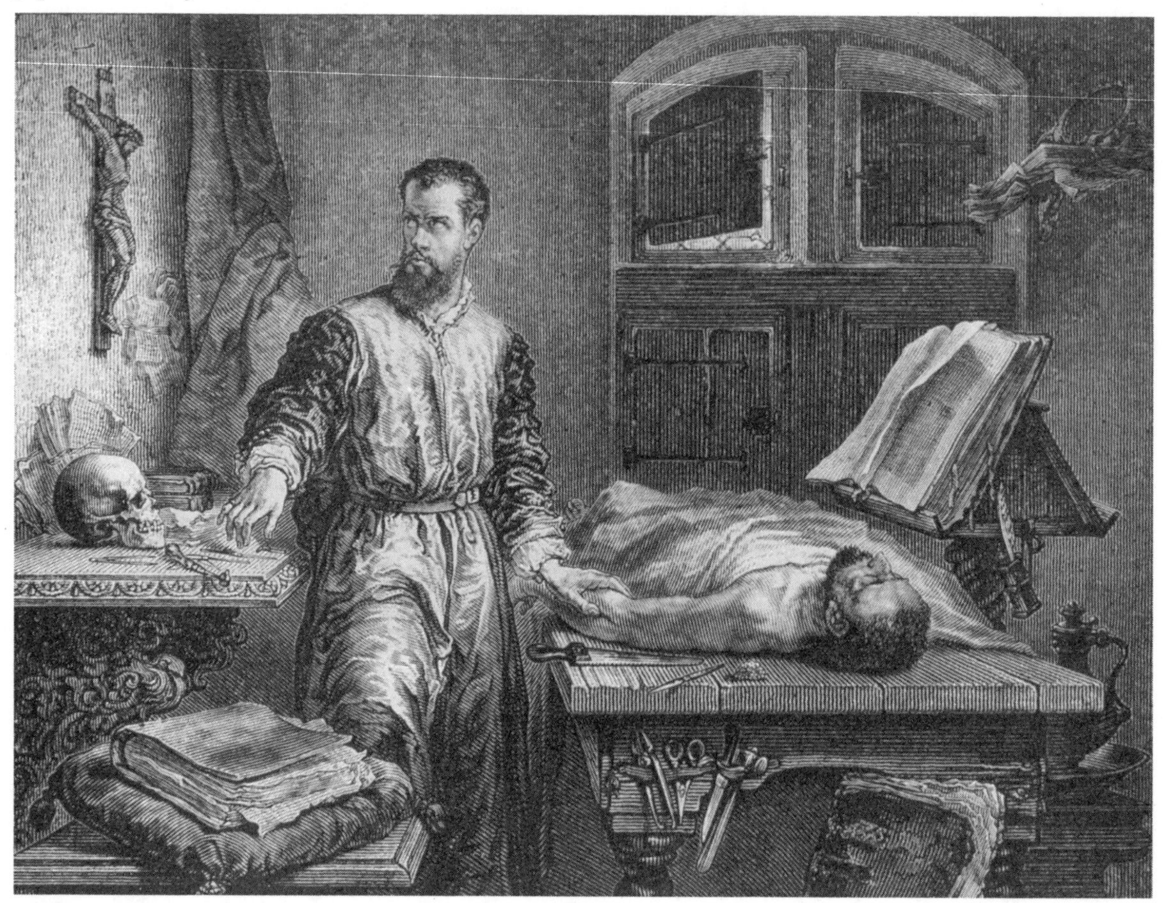

said that the human body must have changed since Galen's time. Vesalius, however, boldly announced that the respected ancient authority was wrong. He later discovered that Galen had gotten most of his information from the bodies of monkeys, dogs, and pigs instead of from human beings.

In 1543, the same year that Copernicus published his revolutionary description of the solar system, Vesalius published an equally revolutionary book describing human anatomy. The book had beautiful illustrations by an artist named Calcar. They showed nerves, muscles, and other parts of the body in great detail.

Vesalius foresaw that the book would bring him trouble. He dedicated it to Charles V, the Holy Roman Emperor—who controlled what are now Germany, Spain, and nearby countries in central Europe—and asked the emperor for protection. "I am aware how little authority my efforts will carry by reason of my youth," Vesalius, who was just twenty-eight at the time, wrote in the book's introduction. He feared "the attacks of those who have not applied themselves to anatomy."[48] Some attacks would be likely to come from other anatomy teachers, who would object to his questioning Galen. More dangerous, though, would be an attack from the church, which objected to most studies of anatomy.

Vesalius was right to worry. The book's publication cost him his job in Padua. He was so persecuted that he eventually gave up the study of anatomy entirely and moved to Spain, where he became court physician to Charles V. Even then the church kept a wary eye on him. He narrowly escaped being put to death for what he had written.

Like Andreas Vesalius, William Harvey (pictured) formulated anatomical theories based on his own observations rather than on the teachings of authorities.

A Living Pump

Vesalius's book remained in circulation. Many medical students, however, went on being taught anatomy just the way they had before it was written: by seeing a few dissections and reading Galen. This was the kind of training a young Englishman named William Harvey received in England's famous Cambridge University at the end of the sixteenth century. Fortunately for science, Harvey, like Vesalius, went to Padua to finish his medical studies. There he had an anatomy teacher named Fabricius, who followed the methods of

Vesalius. From him Harvey learned to trust his own eyes instead of the words of authorities.

Harvey returned to England and became a successful London physician. In time he became the doctor and close friend of Great Britain's king, Charles I. He never lost his interest in anatomy, however. He made many dissections of both human and animal bodies. On the basis of these dissections, he decided that Galen was often wrong about how the body worked as well as how it was built.

The heart and blood were Harvey's special interest. Galen had believed that the body had two kinds of blood. Dark blood was carried in blood vessels, or tubes, called veins, and bright red blood flowed in other blood vessels called arteries. The liver made dark blood from nutrients in food. Red blood came from the heart, which heated it. Galen thought blood flowed into the body from the liver and heart and stayed there until it was used up.

By cutting open animals whose hearts were still beating, Harvey discovered that the heart was not a heater but a pump. It was much like the water pumps that were beginning to be used in England. The heart was made of muscle, and it contracted, or squeezed together, and relaxed just like muscles in an arm or leg. Each heartbeat, or pulse, was a cycle of relaxing and contracting that pushed blood through the body.

Harvey described the anatomy of the heart in a book published in 1628, titled *An Anatomical Essay on the Motion of the Heart and Blood in Animals*. He showed how blood moved from the heart's two upper chambers, or atria, to its two lower chambers, or ventricles, and that from there it flowed into the body through the arteries.

Blood Flows in Circles

The idea of the heart as a pump was startling enough, but the second major idea in Harvey's book was even more so. Harvey claimed that instead of being constantly used up and re-created,

> the blood in the animal body is impelled [pushed] in a circle. . . . This is the act or function which the heart performs by means of its pulse. . . . It is the . . . only end [purpose] of the motion and contraction of the heart.[49]

Indeed, Harvey said, blood flowed in two circles. One took the blood through the lungs. The other carried it through the rest of the body. The blood flowed away from the heart in the arteries and returned to the heart in the veins.

Harvey showed that the veins contained doorlike valves, which opened in only one direction. Because of these valves, blood in the veins could flow only toward the heart. This observation disproved Galen's belief that blood flowed in the same direction in both arteries and veins.

Harvey also supported his ideas by measuring the amount of blood the heart pumped in half an hour. He showed that this amount was more than the whole body could hold. This suggested that the blood must be pumped through the heart over and over rather than being newly made each time.

A few other anatomists had suggested that blood moved in a circle between the heart and lungs. None, however, had proposed that the blood also circulated throughout the body. Harvey wrote that this last idea was "so . . . unheard-of that I not only fear injury to myself from the envy of a few, but I tremble lest I have mankind at large [as a whole] for my enemies"[50] because of it.

Harvey was luckier than Vesalius, however. He was not severely criticized, and his ideas slowly seeped into European medical thinking. Within twenty years Harvey could write to one of his few remaining critics, "I perceive that the wonderful circulation of the blood, first found out by me, is consented to [agreed with] by almost all."[51]

Harvey was unable to answer only one major question about the circulation: how the blood traveled from the arteries to the veins. To discover this missing link he needed a microscope. Compound micro-

William Harvey was fascinated by the anatomy of the heart. By performing vivisection (pictured), he discovered that the heart acts as a pump to circulate blood throughout the body.

scopes—those with two lenses—began to be used in Europe around 1590. They were not common in Harvey's time, however, and he did not use one.

In 1660, three years after Harvey's death, an Italian scientist named Marcello Malpighi, using a microscope, discovered tiny blood vessels that connected the arteries and veins in the lungs of a frog. These vessels came to be called capillaries. Malpighi showed that, as Harvey had predicted, blood flowed in only one direction in the capillaries—from arteries to veins.

From Malpighi's time on, microscopes opened up increasingly detailed views of the hidden world within the bodies of living things. In 1665 Robert Hooke, a friend of chemist Robert Boyle, described microscopic walled structures in cork, which was made from tree bark. Hooke called these structures cells because they reminded him of the tiny, bare rooms, called cells, in which monks lived. He noted that under the microscope a thin slice of cork looked "all perforated and porous, much like a Honey-comb. . . . These pores, or cells, . . . consisted of a great many little Boxes."[52] In fact, Hooke was seeing only the thick walls that remained after the living cells of the tree had decayed.

When more powerful and accurate microscopes were developed in the 1820s, scientists became able to see many kinds of cells in living things. In 1838 a German scientist, Matthias Schleiden, proposed that cells were the basic living units of which all plants were made. A year later a second German, Theodor Schwann, said the same was true of animals. "There is one universal principle of development for the elementary parts of organisms however different, and . . . this principle is the formation of cells,"[53] Schwann wrote. He compared cells to the bricks used to build a house. A later German anatomist, Rudolf Virchow, wrote that a living body is a "state of which every cell is a citizen."[54]

"Little Animals"

At the same time that microscopes revealed the hidden world of the body, they also exposed a second and even stranger world: that of single-celled microorganisms. The first to peer into this world was Antonie van Leeuwenhoek, a Dutch cloth merchant. Leeuwenhoek did not use the compound microscopes that Hooke and Malpighi had employed. Instead, he made his own simple, or single-lens, microscopes. He ground the lenses so carefully

An illustration of Robert Hooke's compound microscope, through which he discovered tiny structures called cells.

Around 1673, Antonie van Leeuwenhoek began making single-lens microscopes. These microscopes allowed Leeuwenhoek to see "little animals"—what are now called single-celled microorganisms.

that they revealed far more than could the crude compound microscopes of his time.

Beginning in 1673 Leeuwenhoek sent letters and drawings describing what he saw to the British Royal Society and other scientific societies. The letters covered everything from cells in blood to mouthparts of honeybees and flies. But the strangest letters were the ones that told about what Leeuwenhoek called "little animals." He found these lively one-celled creatures everywhere—in rainwater, in beer, in matter that he scraped from his teeth. He was amazed at the creatures' great number. "All the people living in our United Netherlands are not so many as the living animals that I carry in my own mouth this very day!"[55] he wrote.

Microbes Are Everywhere

As with cells, a true understanding of Leeuwenhoek's "little animals" had to wait for the better microscopes of the nineteenth century. It also had to wait for the brilliant mind of a French scientist, Louis Pasteur.

Pasteur did not set out to study living things at all. He had trained as a chemist. But while he was teaching in the French industrial city of Lille, a factory owner asked him to find out what was spoiling the alcohol that the factory made from beet juice. This request turned Pasteur's attention to fermentation, the natural process by which beer, wine, and other products containing alcohol are made.

In a breakthrough discovery, Louis Pasteur proved that living microorganisms called yeast were responsible for fermentation. He also showed that certain microbes could cause meats and beverages to spoil.

Little Animals in the Mouth

In a letter to the Royal Society dated September 17, 1683, Antonie van Leeuwenhoek described the microorganisms he saw in matter scraped from his teeth. Leeuwenhoek's letter is quoted by René Dubos in The Unseen World, *a book about discoveries made through the microscope.*

"I . . . saw, with great wonder, that in the . . . [scraped] matter there were many very little living animalcules [microbes], very prettily a-moving. The biggest sort . . . had a very strong and swift motion, and shot through the water (or spittle) like a pike [a kind of fish]. . . . These were most always few in number.

The second sort . . . oft-times spun round like a top, and every now and then took a [zigzag] course . . . and these were far more in number.

The third sort . . . at times . . . seemed to be oblong [shaped like a flattened circle], while anon [at other times] they looked perfectly round. . . . They went ahead so nimbly, and hovered so together, that you might imagine them to be a big swarm of gnats or flies, flying in and out among one another."

Most scientists of the time believed that fermentation was strictly a chemical process. Pasteur, however, proved that it was carried out by living microorganisms called yeast. It is a process by which yeast and some other microorganisms break down food to gain nutrients. Pasteur also showed that putrefaction, or decay, is another breakdown process, carried out by different microorganisms. He discovered particular microbes, as they came to be called, that make meat rot, milk go sour, and wine spoil.

To find out how widespread microorganisms were, Pasteur traveled to locations ranging from Paris basements to high peaks in the Alps during the summer of 1860. He brought with him flasks of a nutrient solution that had been boiled to kill all microbes inside and then sealed shut. The broth would remain clear for years if the flasks stayed sealed. Whenever Pasteur opened the flasks, microbes almost always grew in them, turning the broth cloudy.

Pasteur found microorganisms in the air everywhere he went. They were most common, however, in places where many people and animals lived. In other experiments Pasteur showed that the microbes lived on dust in the air rather than in the air itself.

Invisible Killers

Pasteur's studies of microbes made him wonder whether these invisibly tiny living things could cause disease. Some diseases were similar to putrefaction, caused by microbes. The presence of microbes everywhere in the air might explain why certain diseases spread so easily from one living thing to another. Pasteur wrote:

> When we see beer and wine subjected to deep alterations because they have given refuge to micro-organisms invisibly introduced and now swarming within them, it is impossible not to be pursued by the thought that similar facts may, *must*, take place in animals and in man.[56]

During the 1870s Pasteur and other scientists, including German physician Robert Koch, showed that microorganisms could cause a variety of diseases in animals and humans. To prove that a particular microbe caused a disease, scientists grew that kind of microbe in a dish or test tube containing nutrient substances. They repeated the process until they were sure they had only one kind of microbe in this artificial colony, or culture. Then they injected a tiny amount of the culture into test animals. If the animals got the disease, that microorganism had to be its cause.

Pasteur hypothesized that airborne microbes could be responsible for spreading human diseases. Pasteur spent many hours in his lab trying to prove his theories.

Doctors in Pasteur's time went from patient to patient in hospitals without washing their hands or instruments. Such behavior, Pasteur saw, was sure to spread disease microbes. He tried to warn the doctors that they were unintentionally killing their patients. "I shall force them to see; they will have to see!"[57] he once shouted to his assistants. Most doctors, however, refused to believe that these invisibly tiny living things could be as dangerous as Pasteur claimed.

Fortunately, a few farsighted doctors did begin using Pasteur's ideas to prevent human disease. Joseph Lister, a British surgeon, did so even before Pasteur himself suggested it. In the early 1860s Lister read Pasteur's research showing that microbes were almost everywhere and that some could cause decay in dead matter. Lister concluded that microbes might also cause the kind of decay that occurred in open wounds caused by accidents or surgical operations. This decay, called gangrene,

Protecting Patients from Microbes

In a lecture given to a group of physicians and surgeons in 1878, Louis Pasteur warned doctors against the danger of disease microbes. He also suggested ways of keeping such microbes out of patients' bodies. René Vallery-Radot, Pasteur's son-in-law, quoted this lecture in his Life of Louis Pasteur.

"The water, the sponge, the charpie [lint] with which you wash or dress a wound, lay on its surface germs which, as you see, have an extreme facility [ease] of propagating [growing] within the tissues, and which would infallibly [surely] bring about the death of the patients within a very short time. . . . If I had the honour of being a surgeon, convinced as I am of the dangers caused by the . . . microbes scattered on the surface of every object, particularly in the hospitals, not only would I use absolutely clean instruments, but, after cleansing my hands with the greatest care and putting them quickly through a flame (an easy thing to do with a little practice), I would only make use of charpie, bandages, and sponges which had previously been raised to a heat of 130° C. to 150° C.; I would only employ water which had been heated to a temperature of 110° C. to 120° C. All that is easy in practice, and, in that way, I should still have to fear the germs suspended in the atmosphere surrounding the bed of the patient; but that is almost insignificant compared to that of those which lie scattered on the surface of objects, or in the clearest ordinary water."

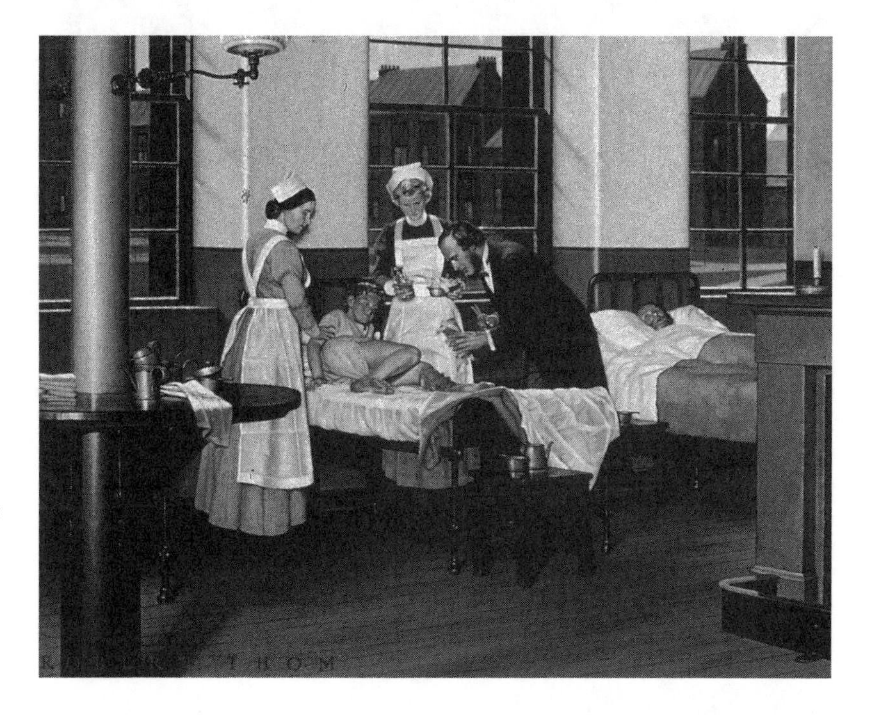

Joseph Lister treats a patient's wounds with an antiseptic dressing. Pasteur's research inspired Lister to use the powerful microbe-killing solution to prevent infection.

often made people lose limbs or even die. In 1864, for example, Lister noted that 45 percent of his patients died of gangrene or other infections—and his success record was better than most.

To try to prevent such deaths, Lister began treating wounds with dressings containing a powerful microbe-killing, or antiseptic, chemical called carbolic acid. Within three years he reduced his patients' death rate after surgery from 45 percent to 15 percent. In 1867 he described his technique in a book called *On the Antiseptic Principle of the Practice of Surgery*.

Vaccination

Keeping microorganisms out of the body by cleanliness and use of antiseptics was one way of preventing disease. A laboratory accident led Louis Pasteur to discover a second way. It happened when he was studying the microbes that caused a bird disease called chicken cholera. Normally these microbes killed most chickens into which they were injected.

In the summer of 1880, just before leaving on a vacation, Pasteur told an assistant to inject some birds with a culture of the cholera microbes. But the assistant, perhaps eager to start his own vacation, forgot. The culture sat on the laboratory shelf for a month. When the assistant returned and did the injections, the injected birds fell ill only briefly and then recovered.

The assistant simply thought the culture had spoiled. Pasteur, however, suspected that something much more interesting had occurred. He injected the recovered chickens with microbes from a fresh cholera culture that killed other chickens very quickly. The chickens that had received the old culture stayed

Edward Jenner vaccinates a young boy against the deadly smallpox disease. Through vaccination, Jenner discovered, a person could be made immune to specific diseases.

healthy. Somehow the older culture had protected them against the disease.

This was not the first time such a thing had happened. Many doctors had noted that if people or animals caught certain diseases and recovered from them, they could not get those diseases again. About eighty-five years before Pasteur began watching his strangely healthy chickens, an English country doctor named Edward Jenner had used this fact to protect people against a deadly disease called smallpox. Jenner noticed that farm women who had previously caught a rash called cowpox did not get smallpox. Cowpox was similar to smallpox but much milder. Jenner then deliberately scratched matter from cowpox sores into people's arms and later exposed them to smallpox. Like Pasteur's chickens, they stayed well. Jenner called his process vaccination, after the Latin word for "cow."

Pasteur suspected that, in effect, his assistant had accidentally vaccinated chickens against chicken cholera. But this time the protective substance, or vaccine, had been made by letting the culture grow older and therefore weaker rather than by using a related but less dangerous microbe. Pasteur later developed vaccines for an animal disease called anthrax and for rabies, a deadly disease that could affect humans.

By the end of the nineteenth century, scientists' exploration of the hidden worlds of human anatomy and microbes was saving lives. Understanding how the body was constructed and how it worked helped doctors treat some kinds of disease. Recognizing that microorganisms could cause disease let doctors apply new methods, such as vaccination and the use of antiseptics, to prevent other illnesses. Reformers in the field of public health also began to clean up filthy living conditions in which microbes could breed. Thanks to these new advances in medicine, deaths from typhoid fever, a common microbe-caused disease, were reduced from 332 per million people in 1871 to only 35 per million by 1911.

Chapter

7 The Wonderful Century

Europeans of the late Middle Ages were sure that, as human beings, they were the most important things in the universe except for God. The Bible explained how God had created humans specially and given them control over all of nature. Everyone knew, too, that the earth was the center of the celestial sphere in which sun, moon, and stars were embedded. Humans might not—perhaps even should not—understand the workings of nature, but they could be sure that God oversaw all those workings and arranged everything for their greatest good.

It was wise to concentrate on God's supreme wisdom and the joys of a heavenly afterlife, because life on earth was often unhappy. Epidemics of the Black Death and other diseases swept through communities, sometimes killing as many as one person in four. Even the richest families shivered in drafty castles and used spices to disguise the taste of food that was often spoiled. Travel and communication were limited to the speed of a fast horse.

By the end of the nineteenth century, however, life was very different. Most people still believed in God, but they were no longer sure how the deity had gone about creating the world or what place humans had in it. They knew that the earth was not even the center of the solar system, let alone the universe. The planet itself was untold millions of years old, far older than humankind. And human beings, so the theory of evolution said, were no different from other animals. Many people thought, incorrectly, that Darwin's theory made them merely latter-day apes.

A Century of Change

But even if they had been no more than that, humans were apes that had come a long way. Indeed, those in the West could see that they had come a long way in just the past few generations. People who went to hospitals now had a reasonable chance of coming out alive. Deaths from some epidemic diseases had been cut drastically. Machines run by steam or, increasingly, electric power had replaced or greatly changed human labor in farm, factory, and household. Farmers raised unusually large crops with artificial fertilizers, and fashionable women sported dresses aglow with artificial dyes. People traveled to other countries by railroad or steamship or sent messages to them instantly by telegraph.

Furthermore, Western science and technology were spreading out to affect the world. Countries such as Great Britain,

During the nineteenth century astonishing advancements were made in science and technology, including the invention of the steam engine (pictured).

France, and Germany were producing more goods than their people could use. They needed new markets for these goods. They also needed more raw materials, such as metals, from which to make them. Seeking both these things, they first explored and then conquered lands all over the world. When Europeans moved into other countries, they brought their technology with them. That technology, ranging from improved weapons to antiseptics and railways, permanently changed the lives of the conquered peoples for both better and worse.

Science and Technology Join Hands

The scientific revolution had begun changing people's attitudes and way of life as early as the Renaissance, when Copernicus dethroned human beings from the center of the solar system and Galileo shortly afterward opened up a new world in the sky. The size and speed of the changes, however, increased greatly in the nineteenth century. For example, Alfred Russel Wallace claimed in an 1899 book called *The Wonderful Century* that twenty-four basic advances in science and technology had occurred in the nineteenth century. According to Wallace, all previous history had produced only fifteen such advances.

One reason for this accelerated change was that science and technology were finally working together. Before the nineteenth century most people interested in probing the secrets of nature did not care whether their discoveries had any practical use. Engineers in the factories that began to appear during the late seventeenth century were equally unconcerned about *why* things worked the way they did. All they wanted to know was *how.* They measured and carried out experiments, but they worked mostly by trial and error rather than being guided by scientific theories.

In the nineteenth century, however, science and technology became closely intertwined. Scientists such as Louis Pasteur cared as much about solving practical problems as they did about studying nature. "There are not two kinds of science—practical and applied," Pasteur said. "There is only Science and the applications of Science, and one is dependent on the other, as the fruit is to the tree."[58] Pasteur made most of his fundamental discoveries while solving problems that threatened French industries.

Increasingly, basic scientific discoveries bore practical fruit. Beginning in Lavoisier's time, new understanding of chemical reactions led to improvements in industrial processes that used chemicals. Late in the nineteenth century this understanding helped chemists create new substances, including dyes, explosives, and plastics. Discovery of the relationship between electricity and magnetism produced electric motors, generators, lights, and more. Conversely, the study of new inventions often led to a better understanding of basic science. Scientists studying steam engines, for example, made fundamental discoveries about how heat could be transformed into motion. These discoveries, in turn, led to improvements in technology or creation of new technologies.

In his 1899 book The Wonderful Century, *Alfred Russel Wallace praised the nineteenth century as having produced more scientific and technological advancements than any previous century.*

The scientific and technological developments of the nineteenth century even created whole new areas of science. These included electrical engineering, biochemistry (the study of chemical reactions in the bodies of living things), and microbiology (the study of microorganisms). Some, such as biochemistry, combined sciences that had formerly been separate.

Science Becomes Popular

Partly because they realized that science was changing their lives so rapidly, people in the nineteenth century, in turn, changed their attitude toward science. For the first time since the scientific revolution began, large numbers of ordinary people became interested in scientific subjects. In the late eighteenth century the famous English writer and thinker Samuel Johnson had complained that "No one ever read a book of science from pure inclination [simply from wanting to]."[59] In the 1840s, however, popular writer Harriet

Martineau said that "the general middle-class [British] public purchased five copies of an expensive work on geology . . . [for every] one [copy] of the most popular novels of the time."[60] People who could not afford to buy science books borrowed them from public libraries.

Lectures were just as popular as books. Audiences for Louis Pasteur's dramatic scientific demonstrations were packed with wealthy and fashionable people. A debate in Great Britain between scientist Thomas Henry Huxley, who supported Darwin's ideas, and the bishop of Oxford, who opposed them, was equally crowded. People discussed the latest findings in science as eagerly as people today might discuss the performance of favorite football or baseball teams.

Interest in Scientific and Technical Education

A growth in science education was both a cause and an effect of people's increased interest in science. In the early 1800s France established a new type of school that focused on scientific and technical subjects. A little later in England, the Mechanics' Institutes were set up to teach such subjects to working people. By 1850 there were six hundred Mechanics' Institutes, with a total membership of over one hundred thousand. At the time, their teaching was said to be "far in advance of the universities of Oxford and Cambridge in regard to the physical sciences."[61]

Science became a popular study during the nineteenth century. Lectures by Louis Pasteur were crowded with people eager to witness his exciting demonstrations.

A Fashionable View of Evolution

In a novel called Tancred, *British author Benjamin Disraeli quotes a fashionable lady's muddled but enthusiastic description of evolution. Disraeli is quoted in* Darwin and the Darwinian Revolution.

"You know, all is development. The principle is perpetually going on. First, there was nothing, then there was something; then—I forget the next—I think there were shells, then fishes; then we [human beings] came—let me see—did we come next? Never mind that; we came at last. And at the next change there will be something very superior to us—something with wings. Ah! that's it: we were fishes, and I believe we shall be crows. . . . You understand, it is all science; it is not like those books in which one says one thing and another the contrary [opposite], and both may be wrong. Everything is proved—by geology, you know."

The new popularity of science also produced changes in scientific societies. Around the 1830s long-established scientific associations such as the Royal Society found themselves challenged by the formation of small scientific societies in different parts of Great Britain. By the end of the century over a hundred such groups existed there. Each group had between one hundred and five hundred members—about as many as the Royal Society had had during the seventeenth and eighteenth centuries. Assuming that the Royal Society in those earlier times had included most British people who were interested in science, science historian Stephen F. Mason says that the growth of the provincial societies suggests that "the number of persons who were actively interested in science [in Great Britain] increased at least a hundredfold during the nineteenth century."[62] Some of these groups combined into new national organizations dedicated to the advancement of science. Their meetings became important places for scientists to exchange information.

Knowledge Is Power

During the nineteenth century governments as well as their citizens began to pay systematic attention to science for the first time. A member of the court that ordered Lavoisier's execution during the French Revolution had proclaimed, "France has no need of savants [scholars]."[63] After the revolution, however, France discovered, as a historian wrote in 1864, that

> everything was wanting [lacking] for the defence of the country—[gun]powder, cannons, provisions [food]. . . . It was exactly those men whose labours had been proscribed [banned] who could give to France what she wanted.[64]

An English scientist demonstrates an electrical experiment for fellow members of the Royal Society. During the 1830s, numerous scientific societies were established in Great Britain; by the turn of the century, there were over one hundred groups in existence.

The French government therefore changed from beheading scientists to supporting them.

Partly because of its government's new attitude, France became the first of the major European powers to excel in science. It was replaced by Great Britain in midcentury and Germany at the century's end. The United States during the nineteenth century gave little support to pure science but produced a river of contributions to technology.

Louis Pasteur once claimed that "science is the highest personification of [a] nation."[65] European governments of his time realized, however, that expertise in science and technology was more than a source of national pride. Scientist and historian René Dubos writes of this period: "Scientific knowledge was becoming a source of wealth. Science had also become essential to the security of the state [nation]. . . . Science had become a necessity to society. . . . Knowledge was power."[66] Discoveries in chemistry, for example, produced both valuable new industries and new explosives that could be used in weapons.

Science Becomes a Profession

Because of the new respect for science that both governments and ordinary people felt, people became able to earn a living in science for the first time. Before the nineteenth century, most scientists did their work as a kind of hobby. Those who worked on science full-time usually either were wealthy, like Boyle and Lavoisier, or were supported by rich patrons. Boyle, for example, paid the publication costs for the *Principia* of his fellow Royal Society member, Isaac Newton, who was not well-to-do. Other scientists earned their living in ways more or less unrelated to science. William Harvey was a medical doctor, for example, and Benjamin Franklin was a printer. Such people did their research in their spare time. Even scientists who worked at universities were paid to teach, not to do experiments. Only in the second half of the nineteenth century did governments, businesses, and universities begin to hire people to carry out scientific research. The word *scientist*, meaning someone who does science as a full-time career, was not coined until 1883.

The late nineteenth century also saw the creation of research institutes, where the new class of professional scientists could carry on their work. The French government, for example, founded the Pasteur Institute in 1888 to provide a place for scientists from all over the world to do the kind of medical research that Pasteur had started. The government provided part of the money to pay for the institute. Most of the rest came from individual donations of thousands of French citizens, whose livelihood or even lives Pasteur had saved.

A Plea for Laboratories

In a letter written to the emperor of France in the late 1860s, Louis Pasteur pleads for government support of research laboratories. René Vallery-Radot quotes Pasteur's letter in his Life of Louis Pasteur.

"If the conquests useful to humanity touch your heart—if you remain confounded [amazed] before the marvels of electric telegraphy, of anesthesia, of the daguerreotype [an early form of photography] and many other admirable discoveries—if you are jealous of the share your country may boast in these wonders—then, I implore you, take some interest in those sacred dwellings . . . described as *laboratories*. . . . They are the temples of the future, of riches and of comfort. There humanity grows greater, better, stronger; there she can learn to read the works of Nature, works of progress and universal harmony."

About the same time, Thomas Edison established a different kind of research institute in Menlo Park, New Jersey. It was dedicated not to pure science but to the creation of inventions that could be sold for profit. It was the first full-time industrial research organization in the United States. Similarly, owners of some large factories in Europe, such as the leaders of the electrical and synthetic dye industries in Germany, set up laboratories for research that would benefit them and their industries. Governments sometimes helped to pay for these private laboratories.

The Dark Side of Science

Some nineteenth-century people were aware that not all the changes brought by science and technology were good ones. The new emphasis on machinery and technology distressed some observers, for example. At midcentury a Swiss writer, Henri Frédéric Amiel, expressed his fear that "the useful will take the place of the beautiful, industry of art, political economy of religion, and arithmetic of poetry."[67]

The growing factories made possible by the new technology brought wealth to their owners but misery to many of their workers. Even children often worked twelve or more hours a day. The burning coal that provided power to steam engines and, worse still, the poisonous by-products of some chemical industries filled the skies with foul-smelling clouds that made watchers think of the underworld.

Governments intent on boosting their power at any cost sometimes put science and technology to even worse uses.

Thomas Edison's laboratory in Menlo Park (pictured) was the first full-time industrial research institute in the United States.

A Hellish Scene

Technology improved nineteenth-century life in many ways, but it also brought problems such as pollution and poor treatment of workers. British author Thomas Carlyle describes a terrible scene at a British iron and coal works. Some of the worst of the abuses Carlyle mentions were cleared up later in the century. Carlyle's letter is quoted in the anthology Pandemonium, *edited by Humphrey Jennings.*

"I was one day [going] through the iron and coal works of this neighbourhood,—a half-frightful scene! . . . A dense cloud of pestilential [poisonous] smoke hangs over it for ever, blackening even the grain that grows upon it; and at night the whole region burns like a volcano spitting fire from a thousand tubes of brick [chimneys]. But oh the wretched hundred and fifty thousand mortals [humans] that grind out their destiny there! In the coal-mines they were literally naked, many of them, all but trousers; black as ravens; plashing [splashing] about among dripping caverns, or scrambling amid heaps of broken mineral. . . . In the iron-mills it was little better: blast-furnaces were roaring like the voice of many whirlwinds all around; the fiery metal was hissing thro' its moulds, or sparkling and spitting under hammers of a monstrous size, which fell like so many little earthquakes . . . and through the whole, half-naked demons pouring with sweat and besmeared with soot were hurrying to and fro. . . . Yet on the whole I am told they are very happy: they make forty shillings or more per week. . . . It is in a spot like this that one sees the sources of British power."

They stockpiled new steel cannons, mass-produced rifles and synthetic explosives designed by the chemists—and tried to justify their preparations for war by citing Darwin. For example, a German general wrote:

> War gives a biologically just [fair] decision, since its decisions rest on the very nature of things. . . . It is not only a biological law, but a moral obligation, and, as such, an indispensable factor in civilization.[68]

In fact, neither Darwin nor his theory supported such a conclusion. Darwin said that nature's survivors were those best suited to their environment, not necessarily those who were strongest or most aggressive.

Science and Progress

In spite of these problems, most people in the nineteenth century felt that science was improving their lives—and that it would go on doing so. Science historian Samuel C. Burchell has written that "for many during the 19th century, science replaced religion and philosophy as a tower of hope and welfare."[69] Some philosophers, such as French thinker Auguste Comte, maintained that scientific thinking should be applied to all areas of human interest.

Many people's admiration for science was closely tied to their belief in progress. Science had taken away people's old reasons for believing that they were the center of the universe, but in return it suggested that there was nothing a determined human mind could not accomplish. People saw Darwin's theory—again, incorrectly—as showing an inevitable advance from lower to higher forms of life. They felt that human history followed the same pattern. People had different ideas about what the next stage of progress would be, but most were sure that science would play an important part in it.

Gregor Mendel's observations about pea plants laid the foundation for the science of genetics. However, the importance of his work was not realized until 1900.

The Brink of a New Revolution

At the end of the nineteenth century, some people thought that science and technology in the century to come would advance mainly by making new discoveries in the fields that had recently opened up. These included synthetic chemicals, prevention of microbe-caused disease, inventions that used electric power, and improvements in transportation and communication.

All these areas would indeed develop further. But in the last years of the century there were also hints that completely new discoveries lay just around the corner. In 1895 a German scientist named Wilhelm Conrad Roentgen found mysterious rays that let him see the bones of his hand as clearly as if skin and muscle had turned to glass. Three months later in France, Henri Becquerel discovered a different kind of radiation that was given off by the heavy element uranium. This form of energy, soon known as radioactivity, would be discovered in certain other elements by scientists such as Marie and Pierre Curie. In 1897, British physicist Sir Joseph Thomson showed that atoms, which John Dalton had thought to be indivisible "wooden balls," in fact contained smaller particles.

Equally unexpected events were taking place in the life sciences. In 1900 several scientists unearthed the forgotten work of a mid-nineteenth-century Austrian monk, Gregor Mendel. Mendel's observations, made while growing pea plants in his monastery garden, began to explain the mechanism behind Darwin's natural selection and started a new science called genetics. Meanwhile, an Austrian physician named Sigmund Freud was having long talks with wealthy women patients who had mental troubles. These chats brought him to some startling conclusions about how the human mind worked.

Most Europeans and Americans of the nineteenth century were well aware of how much they owed to the scientific revolution. They did not realize, however, that they were on the edge of a new scientific revolution that would have even greater effects than the first. Those effects would include atomic bombs, computers, genetic engineering, and the exploration of outer space.

Notes

Chapter 1: The Birth of Modern Science

1. Augustine, *De genesi ad litteram.* In Alan L. Mackay, ed., *Scientific Quotations.* New York: Crane, Russak, 1977, p. 10.
2. Quoted in Philip Cane, *Giants of Science.* New York: Pyramid Publications, 1961, p. 32.
3. Quoted in Robert B. Downs, *Landmarks in Science.* Littleton, CO: Libraries Unlimited, 1982, p. 66.
4. Modern version of Francis Bacon's *Instauratio magna,* in Carole Collier Frick, *The Scientific Revolution: A Unit of Study for Grades 7–10.* Los Angeles: National Center for the Study of History in the Schools, 1992, p. 52.
5. Quoted in Allen G. Debus, *Man and Nature in the Renaissance.* Cambridge, England: Cambridge University Press, 1978, p. 1.

Chapter 2: The Workings of the Universe

6. Quoted in Cane, *Giants of Science,* p. 51.
7. Quoted in Downs, *Landmarks,* pp. 72–73.
8. Quoted in Downs, *Landmarks,* p. 73.
9. Quoted in Dagobert D. Runes, *A Treasury of World Science.* New York: Philosophical Library, 1962, p. 102.
10. Quoted in Downs, *Landmarks,* p. 101.
11. Quoted in Stillman Drake, ed. and tr., *Discoveries and Opinions of Galileo.* New York: Doubleday/Anchor Books, 1957, pp. 27–28.
12. Quoted in Drake, *Discoveries and Opinions,* pp. 182–183.
13. Quoted in Timothy Ferris, *Coming of Age in the Milky Way.* New York: William Morrow, 1988, p. 107.
14. Quoted in Jack Meadows, *The Great Scientists.* New York: Oxford University Press, 1987, pp. 72–73.
15. Quoted in Cane, *Giants of Science,* p. 79.
16. Quoted in Mackay, *Scientific Quotations,* p. 122.
17. Quoted in Downs, *Landmarks,* p. 148.
18. A. Wolf, ed. and tr., *The Correspondence of Spinoza.* London: Allen and Unwin, 1928, p. 80.

Chapter 3: Striking Sparks

19. Quoted in Runes, *A Treasury,* pp. 381–382.
20. Quoted in Edward Tatnall Canby, *A History of Electricity.* New York: Hawthorne Books, 1968, p. 21.
21. Quoted in I. Bernard Cohen, *Benjamin Franklin's Science.* Cambridge, MA: Harvard University Press, 1990, p. 89.
22. Quoted in Canby, *A History,* p. 22.

Chapter 4: Nature's Building Blocks

23. Quoted in William H. Brock, *The Norton History of Chemistry.* New York: W. W. Norton, 1993, p. 68.
24. Quoted in Brock, *The Norton History,* p. 67.

25. Quoted in Brock, *The Norton History*, p. 87.

26. Quoted in L. Pearce Williams and Henry John Steffens, eds., *The History of Science in Western Civilization*, vol. 3, *Modern Science, 1700–1900*. Washington, DC: University Press of America, 1978, p. 127.

27. Quoted in Cane, *Giants of Science*, p. 122.

28. Quoted in Williams and Steffens, *The History of Science*, p. 125.

29. Quoted in Brock, *The Norton History*, p. 104.

30. Quoted in Brock, *The Norton History*, p. 123.

31. Quoted in Cane, *Giants of Science*, p. 144.

32. Quoted in William A. Tilden, *Famous Chemists: The Men and Their Work*. New York: Books for Libraries Press, 1968, pp. 110–111.

33. Quoted in Brock, *The Norton History*, p. 128.

34. Quoted in Williams and Steffens, *The History of Science*, p. 285.

35. Quoted in Mackay, *Scientific Quotations*, p. 164.

Chapter 5: A Changing Earth

36. Quoted in A. Hallam, *Great Geological Controversies*. Oxford: Oxford University Press, 1983, p. 22.

37. Quoted in Hallam, *Great Geological Controversies*, pp. 32–33.

38. Quoted in Ruth Moore, *The Earth We Live On*. New York: Knopf, 1971, pp. 113–114.

39. Quoted in Moore, *The Earth We Live On*, p. 128.

40. Quoted in Moore, *The Earth We Live On*, p. 170.

41. Quoted in Williams and Steffens, *The History of Science*, p. 323.

42. Quoted in Stephen F. Mason, *A History of the Sciences*. New York: Collier Books, 1962, p. 415.

43. Quoted in Mason, *A History*, pp. 413–414.

44. Quoted in Mason, *A History*, p. 416.

45. Quoted in Mason, *A History*, p. 419.

46. Quoted in Gertrude Himmelfarb, *Darwin and the Darwinian Revolution*. Garden City, NY: Doubleday/Anchor Books, 1959, p. 292.

47. Quoted in Meadows, *The Great Scientists*, p. 167.

Chapter 6: Hidden Worlds

48. Quoted in Cane, *Giants of Science*, p. 47.

49. William Harvey, *An Anatomical Essay on the Motion of the Heart and Blood in Animals*. Tr. Robert Willis, rev. Alexander Bowie. In The Harvard Classics, vol. 38, *Scientific Papers*. New York: Collier, 1910, p. 129.

50. Quoted in Harvey, *An Anatomical Essay*, p. 106.

51. Quoted in Geoffrey Keynes, *The Life of William Harvey*. Oxford: Clarendon Press, 1966, p. 322.

52. Quoted in M. D. Anderson, *Through the Microscope*. Garden City, NY: Natural History Press/Doubleday, 1965, p. 35.

53. Quoted in Mason, *A History*, p. 389.

54. Quoted in Mason, *A History*, p. 391.

55. Quoted in Gail B. Stewart, *Microscopes*. San Diego: Lucent Books, 1992, p. 41.

56. Quoted in René Vallery-Radot, *The Life of Louis Pasteur*. Tr. R. L. Devonshire. Garden City, NY: Garden City Publishing 1927, p. 214.

57. Quoted in Vallery-Radot, *The Life of Louis Pasteur*, p. 292.

Chapter 7: The Wonderful Century

58. Quoted in René Dubos, *The Unseen World.* New York: Rockfeller Institute Press/Oxford University Press, 1962, p. 37.
59. Quoted in Himmelfarb, *Darwin*, p. 244.
60. Quoted in Mason, *A History*, p. 411.
61. Quoted in Mason, *A History*, p. 441.
62. Mason, *A History*, pp. 439–440.
63. Quoted in Mason, *A History*, p. 436.
64. Quoted in Mason, *A History*, p. 436.
65. Quoted in René Dubos, *Louis Pasteur: Free Lance of Science.* New York: Charles Scribner's Sons, 1976, p. 85.
66. Dubos, *Louis Pasteur*, p. 10.
67. Quoted in Samuel C. Burchell, *The Age of Progress.* New York: Time-Life Books, 1966, p. 16.
68. Quoted in Himmelfarb, *Darwin*, p. 394.
69. Burchell, *The Age of Progress*, p. 29.

Glossary

accelerate: Make something go faster and faster.

acid: A sour-tasting substance that is one of the major classes of chemical compounds.

Age of Reason: The period from about A.D. 1600 to 1800, so called because the philosophy of the time emphasized reason and logical thinking; also known as the Enlightenment.

alchemy: An early form of chemistry, in which science was blended with religion and magic.

algebra: A type of mathematics that allows mathematicians to represent quantities symbolically and show how they are related to one another.

alizarin: The first dye to be synthesized by deliberately using knowledge of organic chemistry.

amber: Ancient tree sap that has hardened to a rocklike texture.

anatomy: The study of the structure of the body.

antiseptic: A chemical that kills microorganisms.

artery: A blood vessel that carries blood away from the heart.

astrologers: People who use the positions of stars and other astronomical bodies to predict the future.

astronomy: The study of stars, planets, and other bodies in space.

atomic weight: The weight or mass of an atom; the atomic weights for different chemical elements are different.

atoms: Extremely tiny, normally indivisible particles of which all matter is made.

atria: The two upper chambers of the heart.

battery: A device that produces electric current from some other form of energy, such as the chemical reaction between two metals soaked in a salt or acid solution.

biochemistry: The study of chemical reactions in the bodies of living things.

blood vessels: Tubes in the body that carry blood.

calculus: A type of mathematics, invented by Isaac Newton, that expresses how one quantity varies in relation to another.

capillaries: Extremely tiny blood vessels that connect arteries to veins.

carbolic acid: A powerful antiseptic used by Joseph Lister and other late-nineteenth-century surgeons to kill microbes in hospitals and on wounds.

cell: The basic living unit of which the bodies of living things are made.

chemical reaction: A reaction in which two elements or compounds combine or break apart to produce other compounds.

chemistry: The study of the way substances combine with and change each other.

circulate: Move in a circle, as the blood does in the body.

coal tar: A black, sticky substance, left after coal gas is extracted from coal, that proved very useful as the starting point for synthesis of organic chemicals.

combustion: Burning, or rapid addition of oxygen to a substance.

compound: A chemical substance made of two or more elements.

compound microscope: A microscope having two or more lenses.

constellation: A group of stars that suggests a picture in the sky.

contract: Squeeze together; muscles contract and relax.

culture: A colony of microorganisms raised in a laboratory.

dissect: Cut apart, especially for the purpose of learning about a body's structure.

element: A chemical substance that cannot be broken down into other substances and is made of a single kind of atom.

ellipse: A flattened circle.

Enlightenment: Another name for the Age of Reason.

epicycle: A smaller circle within a planet's orbit, according to the ancient astronomer Ptolemy.

evolution: Process of slow change by which types of plants and animals become better adapted to their environment.

experiment: A small, planned change in nature made to test an idea or hypothesis.

extinction: Complete dying out of a type, or species, of plant or animal.

fermentation: A process by which certain microorganisms break down some kinds of nonliving matter to gain nutrients.

fossil: A body or body part of an ancient living thing that has been changed to stone.

friction: Resistance to force that is caused when two substances come into contact.

gangrene: A kind of decay, caused by certain microorganisms, that can occur in open wounds.

generator: A device that produces electric current from mechanical energy, such as a turning magnet.

genus: A group of species of living things that are closely related to each other.

gravity: The force of attraction between two objects, which depends on their mass and the distance between them.

heliocentric theory: The theory of Copernicus that explained that the sun is the center of the solar system, circled by earth and the other planets.

heresy: Disagreement with established religious teachings.

humanism: The belief that human beings and life on earth are more important than religion and an afterlife.

hypothesis: A suggested explanation for something that occurs in nature.

inertia: Tendency of objects to keep doing what they have been doing.

Leyden jar: A jar in which an electrical charge could be stored for later use.

lightning rod: A rod mounted on top of a building that conducts lightning harmlessly into the ground.

lodestone: A piece of iron ore that has become magnetized, or able to act as a magnet.

logic: Orderly reasoning; rules for telling whether a statement is true or false.

mass: The amount of matter in an object; often the same as weight.

mauve: A pale purple dye, synthesized from a coal tar compound, that became popular in the late nineteenth century.

microbe: Same as microorganism.

microbiology: The study of microorganisms.

microorganism: A living thing, usually consisting of a single cell, that is too small to see without a microscope.

Middle Ages: The period in European history lasting approximately from A.D. 450 to 1400.

molecule: A combination of atoms that are joined together; the smallest unit of a compound.

Neptunism: The belief, held by some geologists in the late eighteenth century, that an ancient sea was the most important factor in shaping the earth.

optics: The study of the behavior of light.

orbit: The path, shaped like an ellipse, that a planet makes around the sun or a moon makes around a planet.

organic chemistry: The study of the chemistry of compounds of the element carbon.

oxygen: An element that is added to substances during burning.

pendulum: A weight suspended from a cord or rod so that it swings back and forth.

periodic table: The table published by Dmitry Mendeleyev in 1869 that arranged elements according to their atomic weights and qualities.

phlogiston: A substance, later shown not to exist, that many eighteenth-century chemists believed was released into the air during burning.

physics: The study of the way matter and energy behave and affect each other.

prism: A wedge-shaped piece of glass or similar material that breaks white light into a rainbow.

progress: The belief that history or evolution is a process of steady change from a simpler or poorer state to a more complex and better one.

pulse: Heartbeat.

putrefaction: Decay, a breakdown process carried on by some kinds of microorganisms.

radioactivity: A kind of radiation given off naturally by certain elements as their atoms break down into those of other elements.

reflector: A kind of telescope in which incoming light hits a mirror and is then bounced to a lens in the eyepiece.

refraction: Bending of light as it passes from one substance to another.

refractor: A kind of telescope in which a large lens captures light and bends it toward a smaller lens in the telescope's eyepiece.

Renaissance: The period in European history between about A.D. 1400 and 1600, so called because of the rebirth, or rediscovery, of ancient learning that took place.

respiration: Exchange of gases in the body of a living thing.

Romanticism: A belief that imagination and feelings are more important than reason.

scientific method: A way of learning about nature in which ideas are tested by observation and experiment.

species: A specific type of living thing, all members of which are closely enough related to be able to mate with each other and produce offspring.

spectrum: A rainbow of colored light, produced by passing white light through a prism.

static electricity: Electricity that exists in objects that have become electrically charged.

synthesize: Make artificially.

technology: Practical inventions.

telegraph: An invention that turns the on-off switching of an electric current into a signal that can be used to send a message.

vaccination: The process of using a vaccine, a relatively harmless material, to stimulate the body's disease-resisting system.

valve: A device that controls the flow of a liquid by opening and closing; valves in the veins open in only one direction, forcing blood in these vessels to flow toward the heart.

veins: Blood vessels that carry blood toward the heart.

ventricles: The two lower chambers of the heart.

Vulcanism: The belief, held by some late-eighteenth-century geologists, that volcanoes were the most important forces shaping the ancient earth.

yeast: A type of microorganism that carries out fermentation, producing alcohol.

For Further Reading

Sanford P. Bordeau, *Volts to Hertz . . . the Rise of Electricity*. Minneapolis: Burgess Publishing, 1982. Lively account of the work of the sixteen scientists whose names became measurements of electricity.

James Burke, *The Day the Universe Changed*. Boston: Little, Brown, 1985. Describes important scientific discoveries, their relationship to each other, and their effect on Western society in lively style.

Philip Cane, *Giants of Science*. New York: Pyramid Publications, 1961. Fifty short biographies of famous scientists. Packs a lot of information into a small space.

Benjamin Farrington, *What Darwin Really Said*. New York: Schocken Books, 1982. Short, clear account of Darwin's most important ideas. Includes quotations from Darwin's works.

Carole Collier Frick, *The Scientific Revolution: A Unit of Study for Grades 7–10*. Los Angeles: National Center for the Study of History in the Schools, 1992. A useful primer on the scientific revolution.

Stephen F. Mason, *A History of the Sciences*. New York: Collier Books, 1962. Concise, readable account of major discoveries in science from ancient times to the present.

Jack Meadows, *The Great Scientists*. New York: Oxford University Press, 1987. Interesting, well-illustrated biographies of twelve great scientists, including Galileo, Harvey, Newton, Lavoisier, Darwin, and Pasteur.

Ruth Moore, *The Earth We Live On*. New York: Knopf, 1971. Lively history of geology.

Hugh W. Salzberg, *From Caveman to Chemist*. Washington, DC: American Chemical Society, 1991. Readable history of chemistry.

Gail B. Stewart, *Microscopes*. San Diego: Lucent Books, 1992. Book for young people on the development of microscopes and discoveries made with microscopes, such as cells and microorganisms.

Lisa Yount, *The Importance of Louis Pasteur*. San Diego: Lucent Books, 1994. Biography of Pasteur for young people, focusing on his scientific achievements and their effect on nineteenth-century society.

———, *The Telescope*. New York: Walker & Co., 1983. History of the telescope and discoveries made through it, from Galileo to the space telescope. For young people.

Works Consulted

M. D. Anderson, *Through the Microscope.* Garden City, NY: Natural History Press/Doubleday, 1965. History of the development of the microscope and what discoveries made through the microscope have meant to society.

William H. Brock, *The Norton History of Chemistry.* New York: W. W. Norton, 1993. Comprehensive, interesting history of chemistry.

Samuel C. Burchell, *The Age of Progress.* New York: Time-Life Books, 1966. Illustrated history of the nineteenth century includes a chapter on the impact of science and technology.

Edward Tatnall Canby, *A History of Electricity.* New York: Hawthorne Books, 1968. Illustrates the key discoveries, experiments, and inventions that led to our modern knowledge of electricity.

I. Bernard Cohen, *Benjamin Franklin's Science.* Cambridge, MA: Harvard University Press, 1990. Shows the background of Franklin's work and his relationship with other scientists studying electricity.

Allen G. Debus, *Man and Nature in the Renaissance.* Cambridge, England: Cambridge University Press, 1978. Describes the development of humanistic ideas and their relationship to the rise of science.

Robert B. Downs, *Landmarks in Science.* Littleton, CO: Libraries Unlimited, 1982. Brief accounts of important scientific discoveries, including some primary source quotations.

Stillman Drake, ed. and tr., *Discoveries and Opinions of Galileo.* New York: Doubleday/Anchor Books, 1957. Collection of Galileo's journals and other writings.

René Dubos, *The Unseen World.* New York: Rockefeller Institute Press/Oxford University Press, 1962. Describes microorganisms and how Leeuwenhoek, Pasteur, and other scientists learned about them.

Timothy Ferris, *Coming of Age in the Milky Way.* New York: William Morrow, 1988. A thoughtful, dramatic, and readable narrative of astronomical discoveries.

H. S. Glasscheib, *The March of Medicine.* New York: G. P. Putnam's Sons, 1964. Describes major discoveries in the history of medicine.

A. Hallam, *Great Geological Controversies.* Oxford, England: Oxford University Press, 1983. Technical account of arguments between Neptunists and Vulcanists, Catastrophists and Uniformitarians, and other controversies in geology.

William Harvey, *An Anatomical Essay on the Motion of the Heart and Blood in Animals.* Tr. Robert Willis, rev. Alexander Bowie. In The Harvard Classics, vol. 38, *Scientific Papers.* New York: Collier, 1910. William Harvey's classic short book, describing the heart and the circulation of the blood.

Alexander Hellemans and Bryan Bunch, *The Timetables of Science.* New York: Simon and Schuster/Touchstone, 1988. A chronology of the most important people and events in the history of science, with overviews on selected topics.

Gertrude Himmelfarb, *Darwin and the Darwinian Revolution.* Garden City, NY: Doubleday/Anchor Books, 1959. Enjoyable and detailed account of Darwin's theories and their impact on nineteenth-century society.

Humphrey Jennings, ed., *Pandemonium.* New York: Macmillan/Free Press, 1985. Anthology of short quotations describing the impact of the Industrial Revolution between 1660 and 1886.

Alan L. Mackay, ed., *Scientific Quotations.* New York: Crane, Russak, 1977. A collection of short quotations from scientists and people who commented about science throughout the centuries.

David E. Nye, *Electrifying America.* Cambridge, MA: MIT Press, 1991. Describes how the use of electric power changed American life.

Colin A. Ronan, *Science: Its History and Development Among the World's Cultures.* New York: Facts On File, 1982. Interesting, well-illustrated history of science.

Dagobert G. Runes, *A Treasury of World Science.* New York: Philosophical Library, 1962. Extensive collection of primary source quotations from scientists throughout history.

Charles Singer, *A History of Scientific Ideas.* New York: Dorset Press, 1959. Older history of science; somewhat difficult to read.

William A. Tilden, *Famous Chemists: The Men and Their Work.* New York: Books for Libraries Press, 1968. Chapters describe the lives and discoveries of Boyle, Lavoisier, Dalton, and others.

René Vallery-Radot, *The Life of Louis Pasteur.* Translated by R. L. Devonshire. Garden City, NY: Garden City Publishing, 1927. Extremely detailed biography of Louis Pasteur by his son-in-law.

L. Pearce Williams and Henry John Steffens, eds., *The History of Science in Western Civilization,* vol. 3, *Modern Science, 1700–1900.* Washington, DC: University Press of America, 1978. Lengthy selections from great works of science. More difficult to read than Runes's collection.

A. Wolf, ed. and tr., *The Correspondence of Spinoza.* London: Allen and Unwin, 1928. A collection of the philosopher's letters.

Index

Picture Credits

Cover photo: SCALA/Art Resource

American Museum of Natural History, Department of Library Services, neg. no. 319712, 67

Archive Photos, 8, 16 (bottom), 21, 50 (top), 52, 53, 57, 59, 65, 86

Art Resource, 28

The Bettmann Archive, 11, 18, 27, 37, 42, 46 (top), 48, 60, 62, 64, 71, 74, 79 (bottom), 81, 87, 94

Giraudon/Art Resource, 12

Library of Congress, 9 (top), 23, 26, 30, 31, 33, 45, 88, 90, 92

National Archives, 46 (bottom)

National Institutes of Health, 39, 73, 79 (top)

North Wind Picture Archives, 22, 54, 77, 84

Parke-Davis, 9 (bottom), 83

Planet Art, 16 (top)

Reuters/Bettmann, 43

Stock Montage, Inc., 17, 24, 32, 35, 36, 41, 75, 78

Courtesy of the University of Minnesota Libraries, 70

About the Authors

Harry Henderson and Lisa Yount are a husband-and-wife team who live with a large library and four cats in El Cerrito, California. Lisa Yount has written educational material for young people for over twenty-five years. Her other books include *Contemporary Women Scientists* (Facts On File) and *The Importance of Louis Pasteur* (Lucent Books). Harry Henderson is a technical writer who specializes in computer programming and communications. He is the author of *Internet How-To* (Waite Group Press). He has also written material for young people, including *The Importance of Stephen Hawking* (Lucent Books). *The Scientific Revolution* is Henderson and Yount's first joint book.